Conference Proceedings of the Society for Experimental Mechanics Series

Series Editor

Kristin B. Zimmerman
Society for Experimental Mechanics, Inc.,
Bethel, USA

The Conference Proceedings of the Society for Experimental Mechanics Series presents early findings and case studies from a wide range of fundamental and applied work across the broad range of fields that comprise Experimental Mechanics. Series volumes follow the principle tracks or focus topics featured in each of the Society's two annual conferences: IMAC, A Conference and Exposition on Structural Dynamics, and the Society's Annual Conference & Exposition and will address critical areas of interest to researchers and design engineers working in all areas of Structural Dynamics, Solid Mechanics and Materials Research.

Karen Kasza • Jonathan B Estrada • Alexander McGhee •
Kunal Mishra • Michael Keller
Editors

Mechanics of Biological Systems and Materials and the Mechanics of Composite, Hybrid & Multifunctional Materials, Vol. 3

Proceedings of the 2025 Annual Conference on Experimental and Applied Mechanics

River Publishers

Published, sold and distributed by:

River Publishers
Broagervej 10
9260 Gistrup
Denmark

www.riverpublishers.com

ISBN 97887-438-0829-9 (Hardback)

ISBN 97887-438-0834-3 (eBook)

https://doi.org/10.13052/97887-438-0829-9

Conference Proceedings of the Society for Experimental Mechanics

An imprint of River Publishers

Preface

Mechanics of Biological Systems and Materials and the Mechanics of Composite, Hybrid & Multifunctional Materials represents one of five volumes of technical papers presented at the 2025 SEM Annual Conference & Exposition on Experimental and Applied Mechanics organized by the Society for Experimental Mechanics and held in Milwaukee, WI, June 2-5, 2025. The complete Proceedings also includes volumes on: *Dynamic Behavior of Materials; Advancement of Optical Methods & Digital Image Correlation in Experimental Mechanics; Fracture, Fatigue, Failure, Damage Evolution and Thermomechanics & Infrared Imaging; and Mechanics of Additive & Advanced Manufacturing, Inverse Methods and Machine Learning.*

Each collection presents early findings from experimental and computational investigations on an important area within Experimental Mechanics, the Mechanics of Biological Systems and Materials, Micro-and Nanomechanics and other experimental and applied mechanics such as research in progress.

The Biological Systems and Materials segment of this volume summarizes the exchange of ideas and information among scientists and engineers involved in the research and analysis of how mechanical loads interact with the structure, properties and function of living organisms and their tissues. The scope includes experimental, imaging, numerical and mathematical techniques and tools spanning various length and time scales. This symposium at the Annual Meeting of the Society for Experimental Mechanics provides a venue where state-of-the-art experimental methods can be leveraged in the study of biological and bio-inspired materials, traumatic brain injury, cell mechanics and biomechanics in general. A major goal of the symposium was for participants to collaborate in the asking of fundamental questions and the development of new techniques to address bio-inspired problems in society, human health, and the natural world. The 2025 Symposium is the 15th International Symposium on the Mechanics of Biological Systems and Materials. The organizers would like to thank all the speakers and staff at SEM for enabling a successful program.

This volume also includes papers presented from the 11th International Symposium on the mechanics of composite, hybrid & multifunctional materials. These papers highlight how the commercial market for composite continues to expand with a wide range of applications from sporting equipment to aerospace vehicles. This growth has been fueled by new material developments, greater understanding of material behaviors, novel design solutions, and improved manufacturing techniques. The broad range of applications and the associated technical challenges require an increasingly multidisciplinary and collaborative approach between the mechanical, chemical, and physical sciences to sustain and enhance the positive impact of composites on the commercial and military sectors.

New materials are being developed from recycled source materials, leading to composites with unique properties and more sustainable sources. Existing materials are also being used in new and critical applications, which requires a deeper understanding of material behaviors and failure mechanisms on multiple length and time scales. In addition, the unique properties of composites present many challenges in manufacturing and in joining with other materials. New testing methods must be developed to characterize the novel composite properties, to evaluate application and product life cycle performance, as well as to evaluate impacts and merits of new manufacturing methods.

This segment of the volume presents early research findings from experimental and computational investigations related to the processing, characterization, and testing of composite, hybrid, and multifunctional materials.

Editors:
Karen Kasza, *Columbia University, NY, USA*; Jonathan B Estrada, *University of Michigan, MI, USA;* Alexander McGhee, *University of Arizona, AZ, USA;* Kunal Mishra, *Corning Incorporated, NY, USA*; Michael Keller, University of Tulsa, OK, USA

Contents

Chapter 1
Review of Recent Results of Onco-Ultrasound-Tripsy on Cancer Cells

Carmine Pappalettere and Giovanni Pappalettera

Abstract In recent years, cancer research has advanced significantly, resulting in better treatments for various tumour types and extended patient life expectancy. Nevertheless, there is still considerable interest in alternative clinical approaches. A major challenge is creating treatments that precisely target tumours while preserving healthy tissue. Research has shown that low-intensity ultrasound can selectively harm tumour cells due to their unique mechanical properties compared to healthy cells. This paves the way for a technique called Onco-Ultrasound-Tripsy, OUT, which involves using ultrasound vibrations to induce selective mechanical damage to tumour cells. Many advancements have been observed in this context and this survey aims to present several results related to the employment of such approach with applications to lymphoma, breast cancer, osteosarcoma and melanoma cells.

Keywords Cancer treatment · Ultrasound · Cellular fatigue · Selective cellular damage · Onco-Ultrasound-Tripsy

Introduction

The use of ultrasound in medicine dates back to 1942 when Karl and Friederich Dussik attempted to adapt industrial techniques for detecting metal flow to visualize intracranial structures in real time [1]. Around the same time, Raimar Pohlman, influenced by Paul Langevin's discovery that fish reacted strongly to ultrasound, developed a physiotherapy protocol at Charité in Berlin [2]. This protocol set a power density limit of 5 W/cm^2 and required continuous transducer movement during treatment.

In 1942, Lynn & Putnam [3] pioneered the use of ultrasound for targeting brain tissue in animals, marking an early milestone in therapeutic ultrasound. Their equipment leveraged the piezoelectric properties of crystals first discovered by Pierre and Jacques Curie sixty years earlier [4]. They also applied Gruetzmacher's technique, which involved grounding the surface of a vibrating quartz plate to create a concave curvature, similar to a mirror, allowing ultrasound waves to be focused on a precise location.

Since then, ultrasound has been utilized in treating various conditions, including bone fractures, osteoporosis, thrombosis, glaucoma, nerve injuries, skin wounds, and cancer either as a standalone therapy or in combination with drugs [5]. Over time, ultrasound technology has seen major advancements, including miniaturization, improved functionality, more sophisticated control systems, and a wider array of probes and actuators to support expanding medical applications [6]. Some treatments have been approved for clinical use, while others remain in development.

Significant progress has been made in tumor treatment using High-Intensity Focused Ultrasound, HIFU [7, 8]. HIFU directs high-energy ultrasound waves to a small target area, preserving the surrounding tissue. The ablated tissue volume typically measures 1–3 mm (transverse) × 8–15 mm (along the beam axis) after a single exposure. The two primary mechanisms behind HIFU-induced tissue damage are mechanical energy conversion into heat and cavitation.

Additionally, low-intensity ultrasound, with drug assistance, has been investigated for anti-tumor applications, including sonodynamic therapy, which combines low-intensity ultrasound with a chemotherapeutic agent (sonosensitizer) [9]. This approach (called LIPUS) holds promise due to its ability to selectively target cancer cells while preserving healthy tissue.

Carmine Pappalettere · Giovanni Pappalettera
Department of Mechanics, Mathematics and Management, Polytechnic University of Bari, 70126 Bari, Italy
e-mail: carmine.pappalettere@poliba.it; giovanni.pappalettera@poliba.it

Karen Kasza et al. (eds.), *Mechanics of Biological Systems and Materials and the Mechanics of Composite, Hybrid & Multifunctional Materials, Vol. 3*, Conference Proceedings of the Society for Experimental Mechanics Series, https://doi.org/10.13052/9788743808299-1

In general, when ultrasound interacts with tissue, it produces three main effects: cavitation, hyperthermia, and resonant excitation [10, 11]. Unlike cavitation and thermal effects, resonant excitation operates based on the different resonance frequencies of healthy and cancerous cells, allowing for targeted therapeutic action.

The so called Onco-Ultrasound-Tripsy approach (OUT) aims to exploit the mechanical properties of cancer cells for selective damage, emphasizing the potential of low-intensity ultrasound to inhibit cavitation and thermal effects while promoting resonant and fatigue ones. Changes in resonance frequency due to modifications in mechanical properties, such as a demonstrated reduction of up to 70% in the rigidity of cancer cells [12], support this approach.

After having had the idea many years ago and having studied since the 90s the possibility of selectively killing tumor cells by exploiting their different resonance frequency compared to healthy cells, Pappalettere and his collaborators from Polytechnic University of Bari began 20 years ago to deepen the research on the topic. Afterwards, experimental activities were conducted on U937 tumor cells of a myelocytic lymphoma at the University of Fukuoka in Japan (2015) [11, 13–15] and on MCF7 breast tumor cells MCF10 cells at the University of Dundee in Scotland (2016) [16–21] and the corresponding healthy MCF10A at the University of Bari (2019) [22]. Then they continued at the Polytechnic of Bari in collaboration with the University of Foggia on MG-63 osteosarcoma cells (2017) [23–25] and with the University of Bari both on the same U937 cells studied in Japan and on the corresponding healthy CD3/CD8+ Lymphocytes cells and on red blood cells (2023) [26, 27]. Finally, recently, experiments have been conducted on HBL melanoma cells.

Some more recent studies on U937 cancer cells have been developed and validated at the California Institute of Technology (CalTech) basing on the hypothesis that cellular damage to low intensity ultrasound exposure doesn't occur in a quasi-static/low cycle fatigue regime (cellular lysis) but in high cycle fatigue regime (apoptosis) (2020) [28, 29] while additional results can be found in [30].

The aim of this review is to provide an overview of major results obtained by this approach addressing new challenges and new open questions.

Materials and Methods

All the analyses have been conducted in vitro by transferring the cells grown in culture into Petri dishes and sonicating them either from the bottom or from the top of the dish, depending on whether the cells were suspended or adherent. The probe to sonicate the cells was placed in direct contact with the dish's surfaces, coupled with gel, and the dish was filled with culture medium to facilitate wave propagation and prevent air exposure.

Ultrasound waves were generated using the SonoPore KTAC-4000 device (NEPA GENE, Chiba, Japan), capable of producing frequencies between 200 kHz and 5 MHz with a fine adjustment of 1 kHz. The device allows for adjustments in pulse repetition frequency (0.5 Hz to 100 Hz), output duty cycle (0% to 100%), exposure duration, and voltage (0 V to 60 V). The voltage setting directly influences the power output, depending on the specific probe connected. A KP-S20 flat probe (NEPA GENE, Chiba, Japan) with a 20 mm emission diameter was used. For cell counting, Trypsin/EDTA was used to detach the cells from the Petri dish, and the TC20 automated cell counter (Bio-Rad, USA) was employed. Trypan blue was applied to selectively stain dead cells by permeabilizing their membranes. Following this procedure, a 100% viability rate was observed for the living cells. Additionally, the TC20 system measured the cell size, which was found to range from 10 to 15 μm in diameter. Cell mortality was calculated by Eq.1

$$\textit{cell mortality} = \frac{\textit{live cells on control-live cells after treatment}}{\textit{live cells on control}}$$

Applications of Onco-Ultrasound-Tripsy (OUT)

Application on Lymphoma Cells

Ultrasound exposure was applied to the U937 human histiocytic lymphoma cell line. The U937 cell line, sourced from ECACC (catalog number 85011440, UK), was originally derived from the malignant cells of a 37-year-old Caucasian male diagnosed with histiocytic lymphoma. This monocytic leukemia model retains key monocytic features, including lysozyme secretion, esterase activity, and limited phagocytic capacity, consistent with its histiocytic lineage. U937 cells, widely classified as monocytic cells, were maintained at 37°C in a humidified 5% CO_2 incubator using RPMI-1640 medium (Euroclone, Italy) supplemented with 10% fetal bovine serum, L-glutamine, penicillin G, and streptomycin. The culture medium was refreshed every 24 hours to sustain a cell density between 5.0 × 105 and 1.0 × 106 cells/mL. Cells were expanded in T25

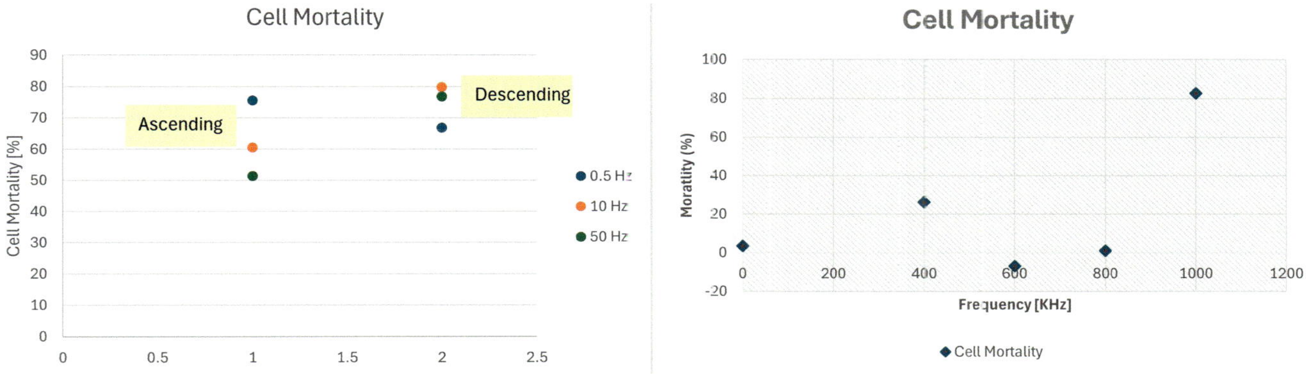

Fig. 1 Analysis of the observed mortality in human leukemic U937 cells was conducted both before (untreated) and immediately after (sonicated) ultrasound treatment. (Left) treatment at sweeping frequencies. (Right) Treatment at fixed frequencies. Results for 0 KHz are referred to untreated cells.

flasks, harvested on the day of the experiment, and resuspended in fresh medium at a concentration of 1 million viable cells/mL. This step was implemented to minimize CO2 accumulation, which may interfere with ultrasound energy propagation and subsequently impact cell mortality outcomes. U937 cells were seeded in 12-well plates at a density of 106 cells/mL in a total volume of 2 mL, with ultrasound exposure applied using a coupling gel.

First evaluations were done by exposing cell to sweeping frequency both in descending and ascending mode [11] by sweeping frequency from 400 kHz to 620 kHz and viceversa at 0.5 Hz, 10 Hz and 50 Hz burst rate.

It was observed that best results are obtained (Fig. 1 left) by working in descending mode at 10 Hz burst rate. In that situation, in fact, 79.8 % of cell mortality was observed. After these treatments the cells that remained alive were put in an incubator for 6 hours and the calculation of the living ones was re-performed. It turned out that after this time the total mortality of the U937 cancer cells rose to about 90% thus showing that the ultrasonic treatment also activated a mechanism of apoptosis of the diseased cells [11].

In a second set of experiments [26, 27] sonication of U937 cells was conducted at constant frequency values of 400, 600, 800, and 1000 kHz, employing voltage levels of 30, 40, 50, and 60 V, a 50% duty cycle, and a 10 Hz burst rate. Exposure durations of 180 and 360 seconds were assessed to identify the optimal conditions for reducing cancer cell viability.

Cell proliferation was monitored for seven days following treatment, with cell counts recorded at 0, 48, 120, and 168 hours. To ensure nutrient availability and prevent any impact on cell growth, the medium was refreshed every 48 hours.

A preliminary series of experiments was performed to assess the effect of frequency variation on U937 leukemic cells. The cells were sonicated for 180 seconds at a constant voltage of 60 V, with a 50% duty cycle and a 10 Hz burst rate, using frequencies of 400, 600, 800, and 1000 kHz (Figure 1 right). Cell mortality, measured immediately after ultrasound exposure, was significantly higher in cells treated with a 1 MHz frequency (82.73% $\pm$ 2.34) compared to untreated controls. Notably, U937 cells exposed to 400 kHz exhibited a mortality rate approximately seven times higher than untreated cells (26.24% vs. 3.50%), indicating that even lower frequencies can negatively impact cell viability.

These results suggest that among the tested sonication conditions, the 1 MHz frequency induces the highest mortality in this tumor cell type. It is important to note that mortality was calculated by Eq.1 and may be influenced by background mortality in the control group. If additional biological factors contribute to control group mortality, this could result in an apparent negative mortality rate, as observed in the case of the 600 kHz treatment.

Application on Breast Cancer Cells

Sonication experiments were conducted using the breast cancer cell line MCF7 [22]. The cells were cultured in plastic T-75 flasks (Sigma-Aldrich, USA) in RPMI 1640 medium (Gibco Thermo Fisher, USA), supplemented with 10% heat-inactivated fetal bovine serum (Sigma-Aldrich, USA) at 37 °C in a humidified incubator with 5% CO_2.

To investigate the effects of intensity and frequency, cell viability after sonication was evaluated using trypan blue dye exclusion tests on macroscopic samples. For this, the cells were plated in 35 mm Petri dishes and incubated for 24 hours before sonication experiments. Prior to the experiments, MCF7 cells had a confluence of over 90%, while MCF10 cells reached around 60% confluence. After sonication at various voltage/frequency settings, the cells were immediately washed with phosphate-buffered saline (PBS) and detached from the dish bottom by treating with 0.250 mL of 0.05% trypsin for 3 minutes. Then, 1.75 mL of medium was added, mixed, followed by the addition of the previous supernatant, PBS, and trypan blue and finally centrifugate at 1200 rpm for 3 mins.

The following sonication conditions were chosen: a sinusoidal wave in pulsed wave mode, with a pulse repetition frequency of 10 Hz, a 50% duty cycle, and 7 fixed frequencies of 400 kHz, 436 kHz, 472 kHz, 510 kHz, 546 kHz, 582 kHz, and 620 kHz. The tests were conducted either at a constant voltage of 60 V or at a constant power of 0.1 W. This power value represents the average output power across the frequency range of 400–620 kHz and also the median power among the 7 sonication frequencies. Results in terms of observed mortality are shown in Fig. 2.

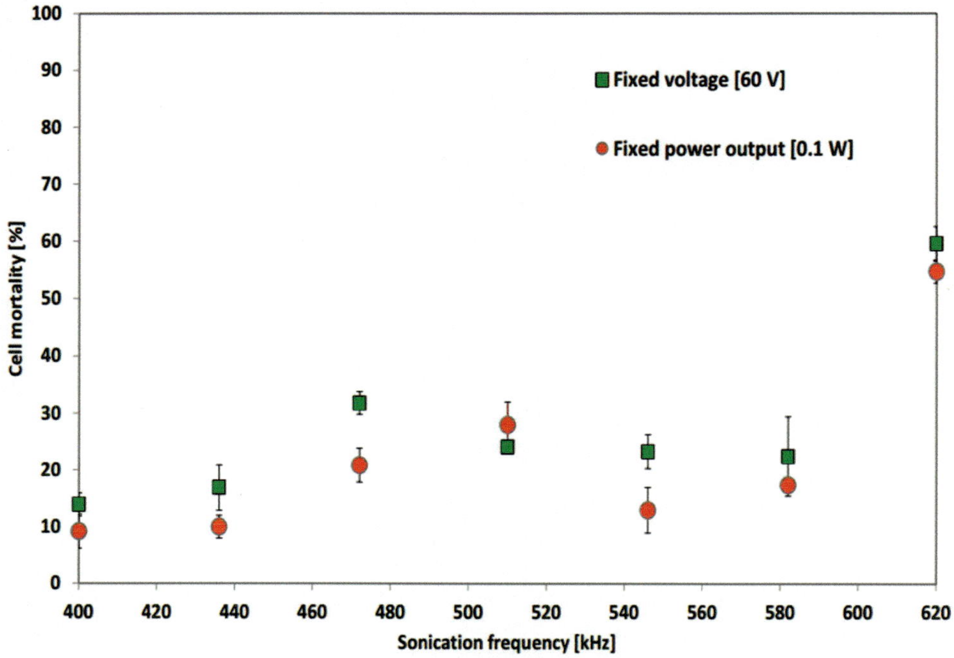

Fig. 2 Response in terms of cell mortality for the fixed voltage condition (green square) and the fixed power condition (red circles)

Figure 2 shows the trend of cell mortality while varying the frequency. The two tested conditions show quite similar trends indicating that cell mortality is mostly correlated with frequency. More specifically cell mortality rises 10% to 15% at 400 kHz to 55% to 60% at 620 kHz. However, this increase is concentrated in the 540–620 kHz frequency range.

Application on Osteosarcoma Cells

The MG-63 osteosarcoma (OS) cell line was cultured in Dulbecco's Modified Eagle Medium (DMEM) supplemented with 10% Fetal Bovine Serum (FBS) [25]. Cells were seeded at a density of 8 × 104 cells per T25 flask and incubated at 37 °C in a standard environment with 95% air and 5% CO2. To detach the cells, Trypsin/EDTA was used, after which the cells were re-plated in Petri dishes and returned to the incubator. Cell confluence was monitored using a Nikon Eclipse Ti-U inverted microscope with a C1 Digital Eclipse modular confocal system.

Cells were exposed for 180 s at a fixed frequency in pulse mode, with a burst rate of 10 Hz, 50% duty cycle and 60 V voltage. In an initial phase, different frequencies were tested to identify the ones able to induce the lowest % of live cancer cells after sonication. Five emission frequencies were analyzed: 400, 510, 582, 800 and 1000 kHz. The power density was, in any case, lower than 5.4 W/cm^2. After these preliminary tests, two frequencies were found associated with the lower percentage of living cells after sonication: 800 and 1000 kHz. The sonication tests for these two frequencies were then repeated on two different days with three replications of the same experimental conditions for each day. Moreover, we the growth index of the cells after ultrasound treatment in real time was analyzed. This was done by using a device able to measure variation in cellular impedance. In fact, an increment of the number of adherent cells results in an increment of the measured impedance and this can be used as an indicator of the cellular growth. These growth indexes were then compared with the growth index of the untreated cells. Experiments were carried out using the xCELLigence RTCA instrument (ACEA Biosciences, Inc., CA, USA), placed in a humidified incubator maintained at 37 °C with 95% air and 5% CO_2.

Cells were exposed for 180 seconds at a fixed frequency in pulse mode, with a burst rate of 10 Hz, a 50% duty cycle, and 60 V voltage. Initially, different frequencies were tested to identify those that caused the lowest percentage of live cancer cells post-sonication. Five emission frequencies were evaluated: 400, 510, 582, 800, and 1000 kHz, with a power density always below 5.4 W/cm^2. After these preliminary tests, two frequencies (800 and 1000 kHz) were found to result in the lowest percentage of surviving cells post-sonication. These two frequencies were then tested again on two separate days, with three replicates of the same experimental conditions for each day. Additionally, the cell growth index was monitored in real time following ultrasound treatment. This was achieved by using a device that measures changes in cellular impedance, where an increase in the number of adherent cells corresponds to a rise in impedance, serving as an indicator of cell growth. These growth indices were then compared to that of untreated cells. The experiments were conducted using the xCELLigence RTCA instrument (ACEA Biosciences, Inc., CA, USA), in a humidified incubator maintained at 37°C with 95% air and 5% CO_2.

Figure 3 (Left) presents the percentage of dead cells following sonication for five different tested frequencies: 400, 510, 582, 800, and 1000 kHz. The evaluation of dead cells was carried out using Trypan blue assays, as dead cells are permeable to the dye and can thus be counted. This procedure was performed immediately after the sonication process, and the number of dead cells was compared with the initial number of treated cells to calculate the percentage shown in Figure 3. The data reveals a general increase in the percentage of dead cells as the frequency rises, particularly for frequencies above 600 kHz. The results at 400 kHz should be considered an outlier and warrant further investigation. The highest cancer cell mortality was observed at the two highest frequencies: 45% of cells were dead after sonication at 800 kHz, and 60% of cells were dead at 1000 kHz.

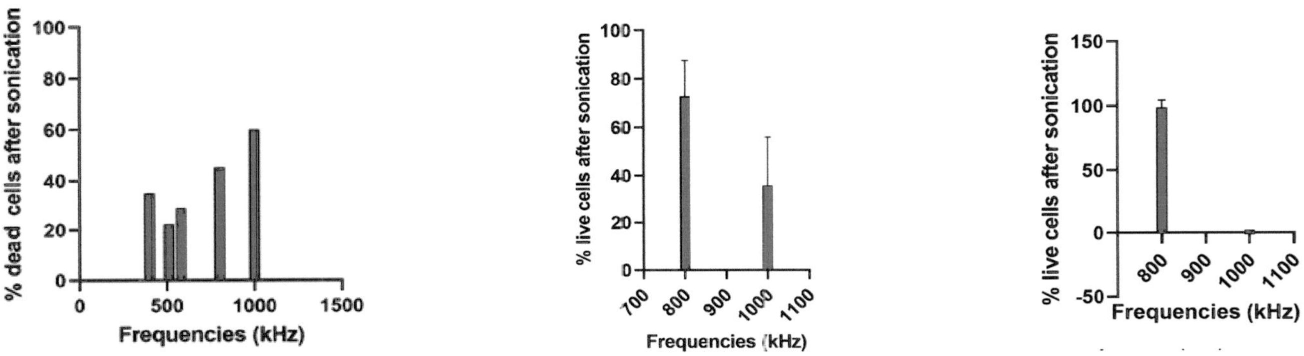

Fig. 3 (Left) Percentage of dead cells after sonication at the five selected frequencies; (Middle) follow-up of the live cells percentage two days after the sonication; (Right) Recalculation of the live cells percentage including considerations about cell/fragment diameter.

The experimental results from the following two days of testing for the selected frequencies of 800 and 1000 kHz are presented in Figure 3 (Middle). The sonication experiments confirm that 1000 kHz is the frequency that results in the lowest percentage of live cancer cells post-sonication, with an average value of under 36%.

Further evaluation was conducted to investigate the possibility that live cells might be present in the culture medium after treatment, potentially detached from the bottom of the dish by the ongoing pressure wave but not damaged by the ultrasound. To rule out this hypothesis, which could result in an overestimation of the number of killed cells, the culture medium was centrifuged to collect the pellet. The pellet was then cultured, and the corresponding growth index was assessed. For all the analyzed dishes, the growth index was found to be zero, indicating the absence of live cells in the medium. Results of this new calculation are presented in Fig. 3 (Right) where analysis is restricted only to cells in the range between 12 and 25 μm.

Application on Melanoma Cells

In order to assess the effects of low-intensity ultrasound on melanoma cells, the HBL metastatic cell line was utilized, known for its high proliferative capacity. These cells were derived from human biopsies at the Department of Translational Biomedicine and Neuroscience (DiBraiN) at the University of Bari "Aldo Moro."

Cells were cultured in Dulbecco's Modified Eagle's Medium High Glucose (DMEM H.G.), supplemented with 10% fetal bovine serum (FBS), 1% L-Glutamine (Euroclone), and 1% Penicillin/Streptomycin (Full DMEM/HG). All cultures were maintained in an incubator at 37 °C with 95% humidity and 5% CO_2.

The culture medium, containing a pH indicator, was refreshed every two to three days based on the requirements of the cell culture. Cells were seeded in four 12-well plates at a density of 100,000 cells per well. These were incubated until reaching a confluence of 70–80%, a prerequisite for initiating ultrasound treatment. Confluence was assessed via optical microscopy. To analyze cell viability and mortality pre- and post-treatment, two control wells were included for each condition.

All plates were exposed to ultrasound using the following sonication parameters: Duty Cycle 50%, Burst Rate 10 Hz, Voltage 40 V, Frequency 1 MHz, and treatment duration of 180 s. These parameters were selected based on preliminary investigations identifying the optimal frequency for inducing cell mortality.

Following treatment, melanoma cells were detached from the wells, washed with 1X PBS, treated with 1X Trypsin-EDTA solution (0.05%/0.02%), and centrifuged at 1100 rpm for 5 minutes.

A cell count was subsequently performed for both control and treated wells using the TC-20 automated cell counter (Bio-Rad) in the presence of Trypan Blue vital dye. Cells were resuspended in 1 mL of complete DMEM/HG medium, and 10 μL of the resulting cell suspension was mixed with 10 μL of Trypan Blue (1:1 dilution). The mixture was then loaded into the counting slide chamber and analyzed using the TC-20 instrument. Cell mortality was evaluated following Eq.1

The findings for HBL melanoma cells are summarized in Figure 4. As depicted, plate #1 exhibited a mortality rate of 100%, while plate #3 and plate #4 showed mortality rates of 83.5% and 85.5%, respectively. However, plate #2 displayed a markedly different outcome, with a significantly lower mortality rate of only 3.33%.

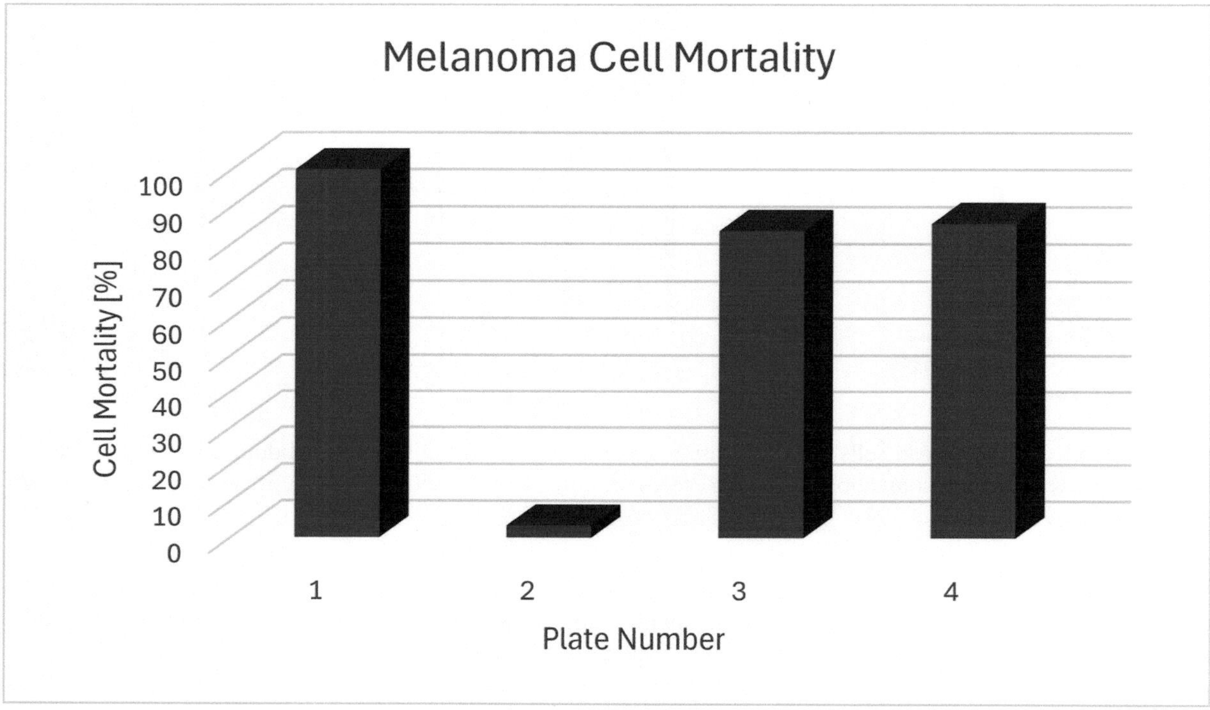

Fig. 4 Evaluation of the cell mortality in the four tested plates

The results obtained for HBL melanoma cells appear somewhat inconsistent and warrant further investigation. While three of the analyzed plates exhibited significant mortality rates, plate #2 showed no such effect. This discrepancy highlights the need for additional studies to identify potential factors underlying these findings.

A closer examination of plate #2's specific conditions is necessary. As described before in this case, cells remained in culture for four days, reaching 100% confluence and displaying a slightly acidified culture medium compared to the previous day. A possible explanation for the lack of effectiveness could be alterations in the culture medium, which also serves as the propagation medium for sound waves. Such changes may have interfered with the resonance regime through absorption and dispersion mechanisms. Furthermore, additional research is required to investigate other potential contributing factors, whether biological or ultrasound-related, that may have influenced the high mortality observed in plates #1, #3, and #4. Ultimately, these findings should be validated through studies with a larger statistical sample.

Conclusion

The overview presented in this paper is testifying the capability of the onco-ultrasound-tripsy approach to face with different kinds of cancer cells at least for what is related to the in-vitro operations. Even if effectiveness of the approach is strictly related to the specific targeted cancer cells in all the presented situation it was found a response in terms of mortality of the cancer cells after exposure. Even more interestingly this response exhibits a frequency dependent behaviour which is fundamental to guarantee selectivity of the approach. Starting from this a lot of efforts are still required to completely understand the whole mechanism of ultrasound-cell interaction, to model them, to understand the role of the cellular micro-environment paving the way of possible preliminary in-vivo investigations. It worth underlining, however, that first evidence of applicability in vivo were shown in [31] where it was observed the capability of the low intensity ultrasound approach in inhibiting cancer growth in mice.

Acknowledgments The Authors would like to thank very much all the researchers who have contributed to the research from The University of Fukuoka (Japan), the University of Dundee (UK), The University of Foggia, The University of Bari and, in particular, from the Polytechnic University of Bari (in alphabetic order): S.N. Barile, C. Casavola, C. Cianci, M.A. Ivone and L. Lamberti.
This work was partly supported by the Italian Ministry of University and Research under the Programme "Department of Excellence" Legge 232/2016 (Grant No. CUP - D93C23000100001)"

References

1. Dussik KT. Weitere Ergebnisse der Ultraschalluntersuchung bei Gehirnerkrankungen. Acta Neurochir (Wien). 1952; 2:379–401.
2. Giugno A, Maugeri R, Graziano F, Gagliardo C, Franzini A, Catalano C, et al. Restoring neurological physiology: The innovative role of high energy MR-guided focused ultrasound (HIMRgFUS). Preliminary data from a new method of lesioning surgery. Acta Neurochir Suppl (Wien). 2017; 124:55–59.
3. Lynn JG, Putnam TJ. Histology of Cerebral Lesions Produced by Focused Ultrasound. American Journal of Pathology. 1944.
4. Manbachi A, Cobbold RSC. Development and application of piezoelectric materials for ultrasound generation and detection. Ultrasound. 2011; 19:187–96.
5. Mitragotri S. Healing sound: The use of ultrasound in drug delivery and other therapeutic applications. Nat Rev Drug Discov. 2005; 4:255–60.
6. Enriquez JL, Wu TS. An introduction to ultrasound equipment and knobology. Crit Care Clin. 2014; 30:25–45.
7. Kennedy JE. High-intensity focused ultrasound in the treatment of solid tumours. Nat Rev Cancer. 2005; 5:321–27.
8. Frassanito MC, Pappalettere C. Physical Principles and Medical Applications of the HIFU Technique (High Intensity Focused Ultrasounds). In Italian. Proceedings of the 38th National Conference of the Italian Association for Stress Analysis. 2009.
9. Son S, Kim JH, Wang X, Zhang C, Yoon SA, Shin J, et al. Multifunctional sonosensitizers in sonodynamic cancer therapy. Chem Soc Rev. 2020; 49:3244–61.
10. Lejbkowicz F, Salzberg S. Distinct sensitivity of normal and malignant cells to ultrasound in vitro. Environ Health Perspect. 1997; 105:1575–78.
11. Ivone M, Pappalettere C, Watanabe A, Tachibana K. Study of cellular response induced by low intensity ultrasound frequency sweep pattern on myelomonocytic lymphoma U937 cells. J Ultrasound. 2016; 19:167–74.
12. Cross SE, Jin YS, Rao J, Gimzewski JK. Nanomechanical analysis of cells from cancer patients. Nat Nanotechnol. 2007; 2:780–83.
13. Ivone M, Pappalettere C, Tachibana K. Effect of Different Ultrasound Frequency Sweep Pattern on Leukemic Cells. 2nd International Conference on Biomedical Engineering and Systems (ICBES'15). Barcelona, Spain: 2015.
14. Ivone M, Pappalettere C, Tachibana K. Study of the effects of low-intensity ultrasound on diseased human monocytes. In Italian. Proceedings of the 38th National Conference of the Italian Association for Stress Analysis. Messina, Italy: 2015.
15. Ivone M, Pappalettere C, Tachibana K. Study of cellular response induced by frequency change of low intensity ultrasound. 2015 E-Health and Bioengineering Conference, EHB 2015. 2016. doi:10.1109/EHB.2015.7391464.
16. Ivone M, Pappalettere C. Biological and mechanical effects of low-intensity ultrasound on suspended and adherent cancer cell cultures. In Italian. 16th National Conference of Clinic Engineering (AIIC). Bari - Italy: 2016.
17. Pappalettere C, Ivone M. Determination of an optimal frequency range for killing cancer cells. In Italian. GIMC-GMA 21st Italian Conference on Computational Mechanics. Lucca -Italy: 2016.
18. Ivone M, Pappalettere C, Tachibana K. Differences in the behavior of attached and floating cells subjected to low intensity ultrasound. ICBMS 2016, IV Conf. on biological and medical science. Budapest: 2016.
19. Ivone M, Pappalettere C, Tachibana K. Differences in the Behavior of Attached and Floating Cells Subjected to Low Intensity Ultrasound. International Journal of Pharma Medicine and Biological Sciences. 2017; 6:7–10.
20. Ivone M, Lamberti L, Pappalettere C. Adherent and suspended cells: different biological and mechanical response to low-intensity ultrasound stimulation. In Italian. Proceedings of the 46th National Conference of the Italian Association for Stress Analysis. Pisa- Italy: 2017.
21. Ivone M, Lamberti L, Pappalettere C. The resonance frequency Effects of low intensity ultrasound on attached and floating cells Viability. ACM International Conference Proceeding Series. 2017:64–68.
22. Ivone M, Lamberti L, Pappalettere C, Caratozzolo M F, Tullo A: Experimental Comparison Of Mcf7 And Mcf10a Response To Low Intensity Ultrasound. J Mech Med Biol. 2019: 19(6) 1950057

23. Cianci C, Lamberti L, Pappalettere C, Ciavarella D, Lo Muzio L. Experimental Evaluation Of Mg-63 Cells Biological Response To Low Intensity Ultrasounds Exposure. COMEC . Brasov: Transilvania University Press Of Bra?ov. 2017:100–03.
24. Pappalettere C, Lamberti L, Cianci C, Lo Muzio L, Illuzzi G, Ciavarella D. Experimental evaluation of the biological response of MG-63 osteosarcoma to low-intensity ultrasound exposure. In Italian. Proceedings of the th National Conference of the Italian Association for Stress Analysis. Villa San Giovanni - Italy: 2018.
25. Pappalettere C, Cianci C, Pappalettera G, Casavola C, Barile SN, Ciavarella D. Onco-ultrasound-tripsy (OUT) treatment for local eradication of MG63 osteosarcoma cell. 13th International Conference on Biomedical Engineering and Technology (ICBET 2023). Tokyo, Japan: 2023.
26. Pappalettere C, Barile S, Cianci C, Pappalettera G, Casavola C. Effects of OUT (Onco-Ultrasound-Tripsy) in-vitro treatment on U937 human histiocytic lymphoma cells. Proceedings of the 52nd National Conference of the Italian Association for Stress Analysis. Genova - Italy: 2023.
27. De Leonardis F, Barile SN, Cianci C, Pisano I, Merla G, Pappalettera G, et al. In Vitro Effects of Low-energy Ultrasound Treatment on Healthy CD3/CD8+ Lymphocytes, Red blood cells, Acute Myeloid leukemia cells, and Jurkat cell line. J Cancer. 2023; 14:1088–106.
28. Schibber EF, Mittelstein DR, Gharib M, Shapiro MG, Lee PP, Ortiz M. A dynamical model of oncotripsy by mechanical cell fatigue: Selective cancer cell ablation by low-intensity pulsed ultrasound. Proceedings of the Royal Society A: Mathematical, Physical and Engineering Sciences. 2020; 476:20190692.
29. Mittelstein DR, Ye J, Schibber EF, Roychoudhury A, Martinez LT, Fekrazad MH, et al. Selective ablation of cancer cells with low intensity pulsed ultrasound. Appl Phys Lett. 2020. doi:10.1063/1.5128627.
30. Pappalettere C, Cianci C, Casavola C, Pappalettera G. Effects of Ultrasound Exposure On Cancer Cells: An In Vitro Study. ICEM20 - 20th International Conference on Experimental Mechanics. Porto: 2023:539–40.
31. Feril L B Jr, Yamaguchi K, Ikeda Dantsuji Y, Furusawa Y, Tabuchi Y, Takasaki I, Ogawa R, Cui ZG, Tachibana K: Low-intensity ultrasound inhibits melanoma cell proliferation in vitro and tumor growth in vivo J Med Ultrasonics 2021; 48:451-461.

Chapter 2
A Bone Remodeling Lab-on-a-Chip to Study Periprosthetic Osteolysis: Early Evidence that Osteocyte Signaling Contributes to Increased Osteoclast Resorption

M. M. Saunders

Abstract Total joint replacement surgeries are highly prevalent throughout the United States. However, it is estimated that within 10 years of surgery, 20% to 30% of patients will present with a complication known as periprosthetic osteolysis (PPOL), or a degradation of the bone surrounding the implant [1]. Patients experiencing PPOL often require revision surgeries, which are associated with higher rates of morbidity and mortality [2]. Thus, there is a great need for therapeutic or pharmaceutical approaches that could delay the need for revision arthroplasties. Currently, the mechanisms that drive this osteolytic process are not fully understood. However, evidence suggests that particulate debris generated from repetitive wear of implant surfaces leads to inflammation within the joint, which ultimately disrupts bone remodeling. Remodeling is a lifelong metabolic process involving three bone cell types (osteocytes, osteoclasts, and osteoblasts) that maintains bone homeostasis by tightly coordinating bone resorption and formation in a dynamic loading environment. Furthermore, mechanical load has also been shown to play a crucial role in bone remodeling and long-term implant stability. The goal of this project was to develop a lab-on-a-chip (LOC) platform that recapitulates the bone dynamic microenvironment and allows for analysis of multicellular signaling processes that lead to PPOL. This *in vitro* system will allow us to study complex signaling between osteocytes, osteoclasts and osteoblasts while eliminating confounding factors present in *in vivo* studies. Following development, the LOC platform was utilized to preliminarily investigate the effects of particulate debris in a dynamic loading environment on osteocyte signaling and the subsequent effect on osteoclast resorption.

Keywords Lab-on-a-Chip · Periprosthetic Osteolysis · Bone Remodeling · Mechanobiology

Introduction

One of the leading causes of late-stage joint failure is aseptic loosening, a condition in which the bond between the bone and the implant fails in the absence of infection. This loosening is often associated with PPOL, or degradation of the bone tissue surrounding the implant. While the mechanisms that drive these processes are not fully understood, many studies have shown that particulate debris generated from wear of the implant surfaces triggers an inflammatory response within the joint, which ultimately leads to bone loss [3–4]. Additionally, long-term implant survival is dependent on the physical activity of the patient. This highlights the crucial role of mechanical load on bone metabolism and long-term implant stability. A more thorough and comprehensive understanding of the underlying mechanisms that regulate load-induced bone remodeling and implant debris-induced osteolysis may lead to novel interventions that mitigate late-stage joint failure.

Background

An investigation of the mechanisms responsible for PPOL requires a deeper understanding of the complex multicellular signaling cascades that regulate bone regeneration. This lifelong metabolic process, known as bone remodeling, is necessary

M. M. Saunders
College of Engineering and Computing, Miami University, Oxford, OH 45056
e-mail: saundem9@miamioh.edu

Karen Kasza et al. (eds.), *Mechanics of Biological Systems and Materials and the Mechanics of Composite, Hybrid & Multifunctional Materials, Vol. 3*, Conference Proceedings of the Society for Experimental Mechanics Series, https://doi.org/10.13052/9788743808299-2

for maintaining the structural integrity of the skeletal system. During this complex multicellular process, old or damaged bone is removed (resorbed) by osteoclasts and new bone is deposited by osteoblasts. A tight coordination between these two processes is required for maintaining overall bone mass. Osteocytes, which account for more than 90% of the total bone cells, play a critical role in this coordination effort. Embedded within the bone matrix, osteocytes exhibit a high degree of interconnectivity and are responsive to external stimuli, such as mechanical load. Once stimulated, osteocytes have been shown to regulate the activity of both osteoclasts and osteoblasts. Furthermore, crosstalk between osteoclasts and osteoblasts has proven essential for coupling the resorption and formation stages of remodeling. While multicellular signaling is crucial to the tight regulation of bone remodeling, many *in vitro* studies are limited to investigating a single, isolated cell type.

In recent years, our lab along with others, has emphasized the need for more complex bone remodeling platforms that more fully recapitulate the intricate bone microenvironment. Our lab has emphasized the benefits of developing a LOC platform for bone remodeling that incorporates osteoclasts, osteoblasts and osteocytes [5]. Here we modified our current platform (Fig. 1) which utilizes fluid shear forces to induce mechanical stimulation to model the periprosthetic microenvironment and investigate cell-cell signaling modalities that lead to PPOL. Per our focus on PPOL and bone resorption, our initial efforts focused upon analyzing the interactions between osteocytes and osteoclasts. The design of our LOC allowed osteocytes to be stimulated via particulate debris under dynamic load while exposing osteoclasts to real-time osteocyte soluble signaling. However, as is typical of novel models, significant effort was required to establish the fidelity of the system. These efforts, which are presented below, included optimizing cell feeding, loading profile and particulate debris dosing. In addition, combined dynamic particulate studies incorporating osteocyte and osteoclast activity were completed with preliminary results provided.

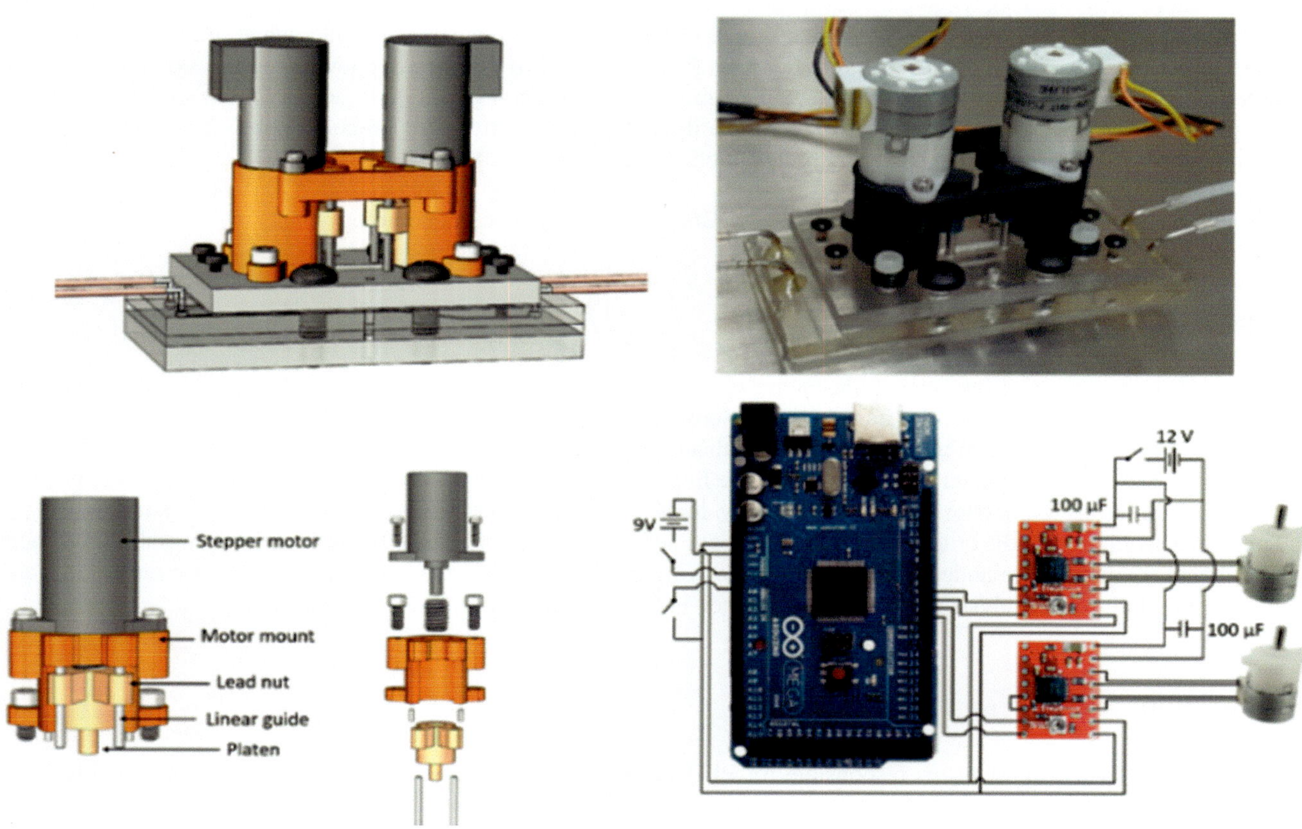

Fig. 1 (top) Schematic of multicellular LOC with pumping platform applied exposing cells to fluid shear forces and photograph of LOC. (bottom) Design of 3D printed linear actuator that used to control fluid flow within the osteocyte channel with exploded view of the linear actuator and design of circuitry used to control movement of stepper motors within the linear actuators.

Analysis

Our previously developed dynamic LOC design incorporating fluid shear was modified to fabricate a multicellular platform that recapitulated the bone microenvironment to serve as a model for PPOL [6]. The modified platform is shown in (Fig. 2). This multilayer system contains two culturing chambers separated by a permeable membrane. In the top culturing chamber, osteocytes are selectively stimulated with particulate debris and mechanical shear stress via oscillating fluid flow. This layer consists of a main cell culturing chamber with a medium reservoir on each end and two sets of access channels. The central access channel is used for filling the medium reservoirs, and the outer access channel is used for seeding cells. The lower culturing chamber contains osteoclasts cultured on bone wafers with a single set of access channels. The two chambers are separated by a thin polyethylene terephthalate (PETE) porous membrane (12 μm thick, 0.4 μm pore size). The design allows osteoclasts to be shielded from the mechanical stimulation and particulate debris present in the osteocyte chamber, while still permitting the real time diffusion of osteocyte soluble signals. Fluid flow within the osteocyte chamber is driven by custom-designed 3D printed linear actuators controlled by an Arduino microcontroller with custom-written software.

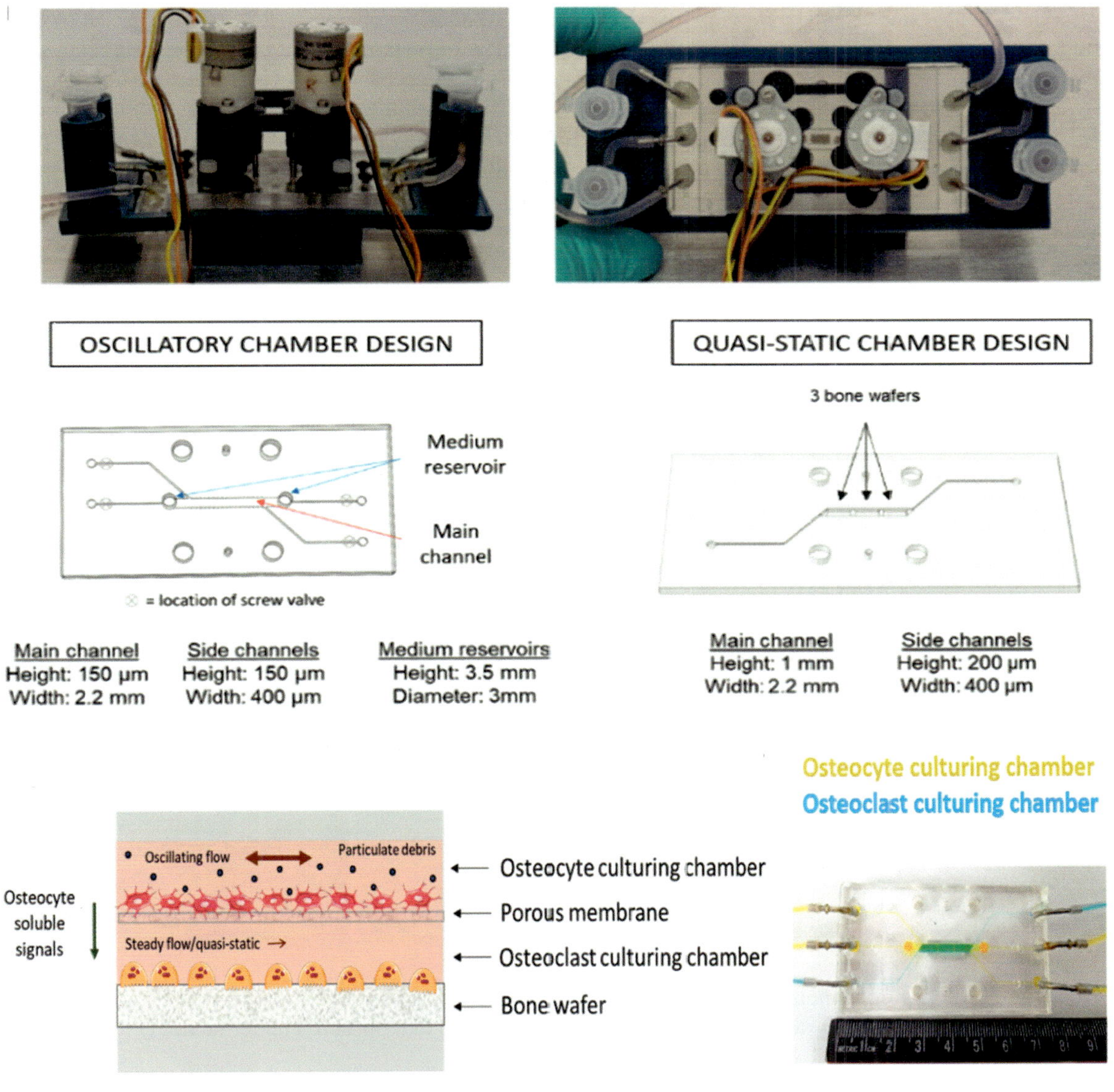

Fig. 2 The assembly utilized for co-culture loading studies quantifying the effects of polyethylene particles in bone loss, modeling PPOL.

To first analyze the effect of mechanical stimulation and particulate debris on osteocytes, only the osteocyte layer of the LOC was used. The fully cured PDMS osteocyte layer was sealed with a PDMS membrane on top and a slab of PDMS (2 mm thick) on the bottom. The main channel of the LOC was coated with collagen type I by flowing the solution through the outside access channels. After 1 h the channel was rinsed with Dulbecco's phosphate-buffered saline. The central access channels were then used to fill the entire chip with Minimum Essential Medium (MEM) α supplemented with 10% fetal bovine serum (FBS) and 1% penicillin/streptomycin.

Optimizing osteocyte feeding in the LOC: MLO-Y4 osteocytes were seeded within the main channel of the chip at a density of 10^4 cells/cm^2 and were given 4 h to adhere to the surface. Assuming a similar culture medium consumption rate as osteocytes grown within a T-25 flask, it was determined that the medium within the main channel should subsequently be replaced with medium from the reservoirs every 3 h. The feeding protocol was fully automated using the linear actuators to drive fluid movement. Chips with a reservoir depth of 3.5 mm contained sufficient culture medium to maintain the cells for 24 h. Every 24 h the medium in the reservoirs was manually replaced. To address the concern of excessive evaporation of culture medium through the PDMS that would result in the formation of air bubbles in the main channel, a humidity chamber was designed and 3D printed to house the chip within the laboratory incubator. This chamber fit on top of a standard 6-well culture plate that was filled with sterile dH_2O. Volumetric flow rates used during the automated portion of the feeding protocol were analyzed to balance osteocyte viability with minimal stimulatory effects. A flow rate of 14.9 µl/h (lower feed rate) was initially tested, which generated a wall shear stress two orders of magnitude less than the lowest value reported in literature that is known to stimulate osteocytes. At this flow rate, osteocyte proliferation rates were substantially less than that of cells grown in a 96-well plate. By increasing the flow rate to 29.8 µl/h (higher feed rate), proliferation rates were comparable to that of cells grown in the 96-well plate (Fig. 3). Additionally, cells displayed typical dendritic morphology as shown for 4 and 72 h (scale bars represent 100 µm); rhodamine phalloidin and DAPI stains at 72 h revealed the presence of an organized cytoskeleton (scale bar represents 50 µm). Based upon osteocytic iNOS production, the higher feed rate was utilized for subsequent studies.

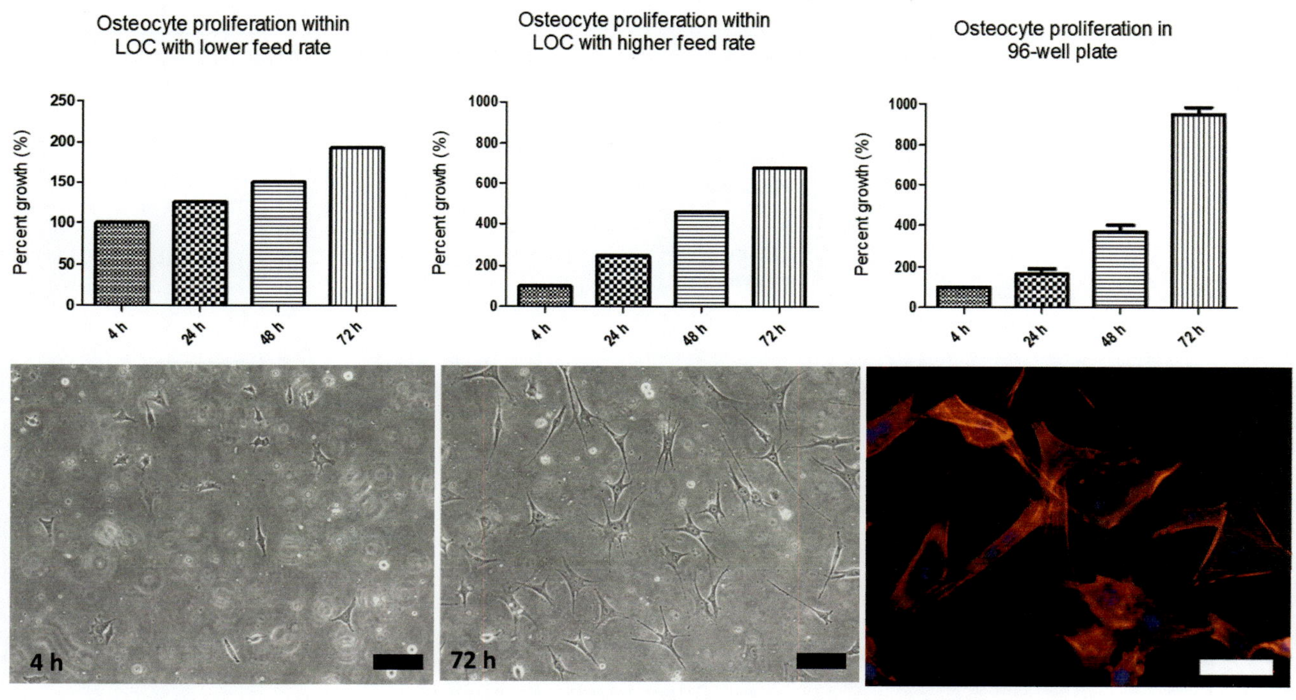

Fig. 3 Osteocytes in the LOC channels maintained viability with cellular proliferation more responsive to a higher feeding rate (29.8 µl/h). At the higher feeding rate, proliferation was comparable to that observed under identical feeding conditions in the 96-well plate.

Validating experimental flow rates: Micro-particle image velocimetry (µPIV) was used to validate fluid flow rates within the osteocyte culture chamber (Fig. 4). A chip was filled with 0.02% solution of fluorescent polystyrene tracer particles with a diameter of 1 μm in minimum essential alpha medium (MEMα) supplemented with 5% fetal bovine serum (FBS), 5% calf serum (CS) and 1% penicillin/streptomycin. The linear actuators were turned on to induce oscillating fluid flow with a sinusoidal wave (1 Hz) and maximum velocity corresponding to a maximum wall shear stress of 1 Pa. Images were obtained using a Zeiss Axio Observer A1 inverted microscope. Burst mode was used to collect 100 images at a rate of 47.6 frames per second. Images were analyzed pairwise using PIVLab, an open-source add-on for MATLAB, and three passes with a decreasing interrogation window (128, 64, and 32 pixels) to generate a velocity vector field. The average particle velocity was determined for each image and then plotted versus time. A one-term sum of sine approximation for the data was obtained using the MATLAB curve fitting application. The amplitude of the model was compared to the theoretical amplitude (previously described), and a calibration coefficient was determined. The calibration coefficient was incorporated into the model, and the experiment was repeated. Osteocytic iNOS production showed that mechanical load continued to have a cellular effect at 24 h. This waveform was used for subsequent loading studies.

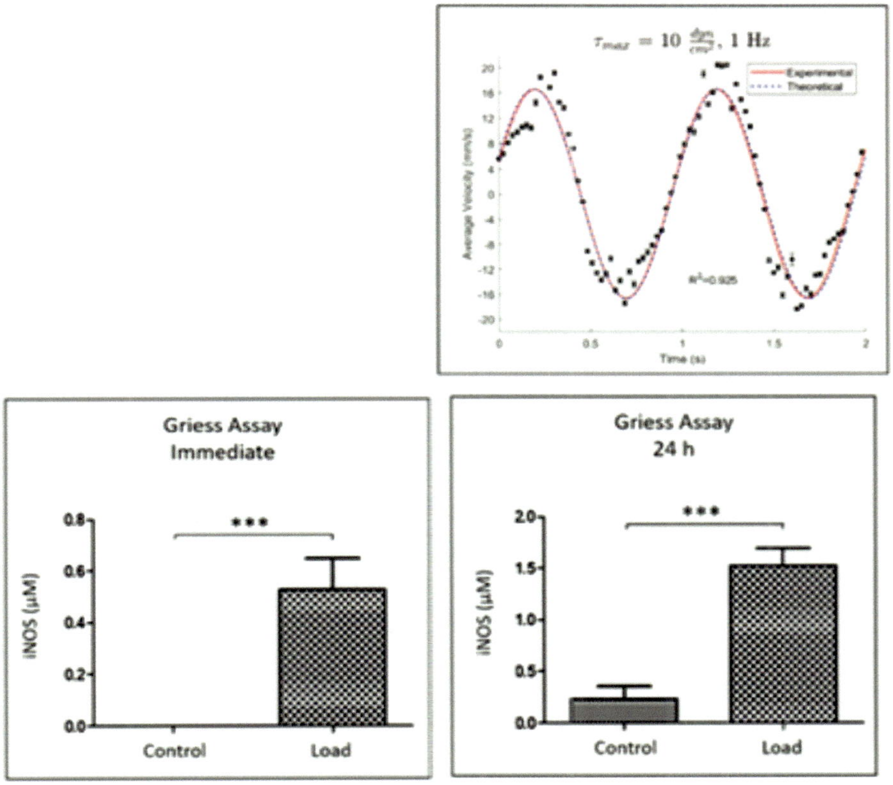

Fig. 4 μPIV analysis to validate fluid flow rate.

<u>Validating experimental particulate debris dosing (particulate preparation):</u> Wear debris was modeled using commercially available polyethylene nanospheres (Cospheric LLC) with a diameter range of 0.2 to 9.9 μm. Evidence suggests that the biological response to debris depends on particle size, and particles with a diameter between 0.1 and 10 μm have been reported to be the most biologically active. A solution of polyethylene particles in culture medium was generated using an established protocol. Briefly, particles were suspended in 70% ethanol then sonicated for 1 h to disperse any clumps. To ensure particles remained in the upper osteocyte chamber, the solution was filtered to remove particles smaller than 0.4 μm. The remaining particles were then resuspended in 70% ethanol, sonicated for 1 h and left at RT for 48 h to eliminate endotoxin contamination. The particles were then washed three times with endotoxin-free PBS, resuspended at 1 mg/ml in MEMα supplemented with 5% FBS, 5% CS and 1% penicillin/streptomycin and sonicated for 30 min. The solution was streaked on nutrient agar plates and incubated for 48 h at 37 °C to confirm the absence of any bacterial or fungal contamination. Prepared PDMS chips were rinsed two times with sterile dH_2O using a 5 ml syringe and Tygon tubing. Using the outside access channels, the osteocyte culture chamber was coated with collagen type I in 0.2 M acetic acid for 1 h then rinsed two times with Dulbecco's phosphate-buffered saline (with calcium and magnesium). The osteocyte chamber and medium reservoirs were then completely filled with MEMα supplemented with 5% FBS, 5% CS and 1% penicillin/streptomycin, taking care to remove

all air bubbles from the system. MLO-Y4 osteocytes were seeded in the chips at 20,000 cells/cm^2 using the outside access channels and a syringe pump set to a speed of 5 ml/h and a volume of 40 µl per chip. The cells were allowed to adhere to the chips for 4 h prior to motor attachment. The linear actuator pump was then used to automatically feed the cells every 3 h with medium from the reservoirs using a flow rate of 29.8 µl/h. Medium in the reservoirs was manually replaced every 48 h. LDH staining and Griess assays were employed to quantify the dose response ranging from 0-1 mg/mL after 48 h. iNOS production was determined using a Griess assay. First, 50 µl of each sample was added to 50 µl of Griess reagent (a solution containing 0.5% sulfanilamide, 0.05% N-(1-Naphthyl)ethylenediamine dihydrochloride and 2.5% phosphoric acid) within a 96-well plate. The absorbance at a wavelength of 540 nm was determined for each sample and converted to a concentration of iNOS. Cells were stained with LDH reaction solution—a mixture containing HBSS, Polypep, diglycine, lactic acid, β-Nicotinamide adenine dinucleotide hydrate and nitroblue 34 tetrazolium. Cells were incubated for 90 min at 5% CO_2 and 37 °C. They were washed with HBSS and fixed in 4% paraformaldehyde for 15 min at RT. They were washed again and stored in HBSS for imaging. Based upon analysis (Fig. 5), particles were resuspended in 1 mg/ml for subsequent experimental studies.

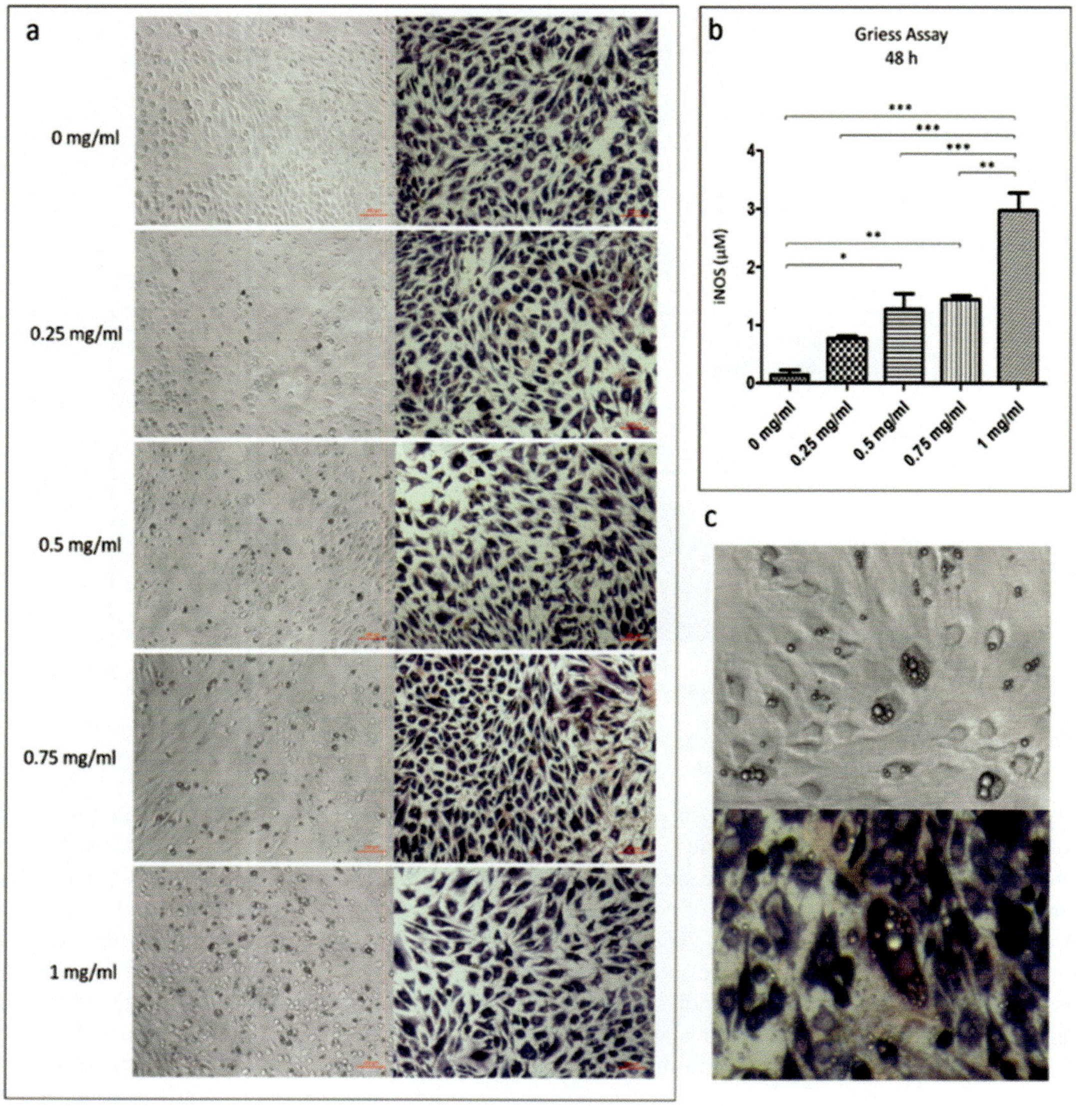

Fig. 5 The results of LDH staining and the Griess assays in quantifying iNOS production as a function of polyethylene particle dose response. Of particular interest are the magnified phase contrast and stained images clearly demonstrating the successful cellular uptake of the polyethylene particles.

Fabricating chips for multicellular studies: Chip fabrication has been previously described. Briefly, a custom-designed mask was designed in SolidWorks 2018 (Dassault Systèmes) and fabricated from Accura SL 5530 using high-resolution stereolithography (Protolabs). Chips were cast from PDMS with a 10:1 ratio of elastomer base to curing agent. The polymer was mixed vigorously, poured over the mask, degassed and cured for 18 h at 45 °C. A similar process was used to fabricate the osteoclast culturing chamber and a thin (0.5 mm thick) PDMS membrane. This membrane served as a diaphragm that, when deflected downward, drove fluid from the medium reservoirs and across the osteocyte culturing chamber. A biopsy punch was used to create holes for the access channels. The PDMS membrane was attached to the top of the osteocyte chamber layer by plasma oxidation for 30 s using a medium RF setting (200 mTorr, 10W). The PETE membrane was attached to the bottom of the osteocyte layer using aminosilanization of the membrane and plasma oxidation. To ensure the PETE membrane was taut, the membrane was initially oversized, and the edges were taped to filter paper that was in turn taped to a flat metal tray. After attaching the membrane to the PDMS layer the excess PETE membrane was trimmed off with a scalpel. The osteoclast culture chamber was then plasma oxidized and pressed to the underside of the PETE membrane following careful alignment with the osteocyte chamber. The chips were then baked at 65 °C for 30 min with weight (~9 N) placed on top. Angled dispensing tips (18-Gauge, 12.7 mm, 90 °) were inserted into the access holes and secured with epoxy.

Analyzing the effects of particulate debris and mechanical stimulation on osteocyte signaling and subsequent osteoclast resorption: The LOC was fabricated with both the osteocyte and osteoclast culturing chambers; experiments were conducted as shown in the experimental design chart (Fig. 6).

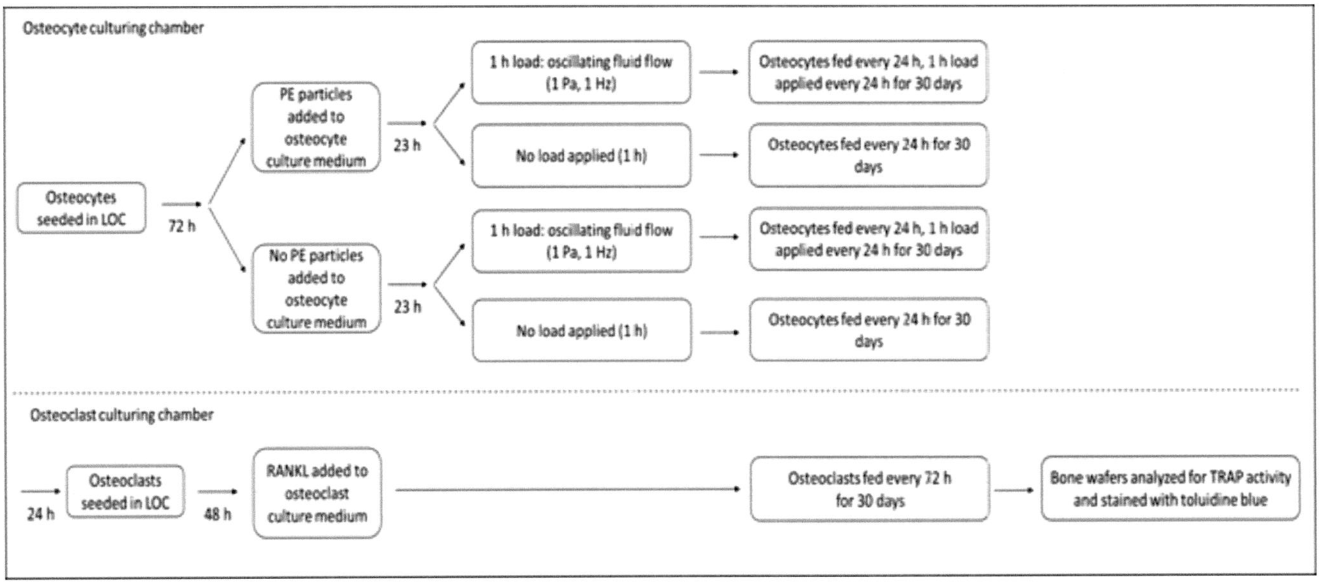

Fig. 6 Experimental design for initial investigation of osteoclast resorption as a function of exposure to osteocytes subjected to mechanical stimulation in the presence of particulate debris simulating PPOL in a dynamic loading environment.

The osteocyte and osteoclast chambers were separated by a porous PETE membrane to allow for diffusion of soluble signals between the culturing chambers. The device was fabricated following the procedures described above. The osteoclast culturing chamber contained three bovine bone wafers (2 mm x 5 mm) secured to the bottom of the channel. Following experimentation, bone wafers were removed from the chip and each wafer was analyzed separately. MLO-Y4 osteocytes were seeded in the upper chamber at a density of 10^4 cells/cm^2 using a flow rate of 5 ml/h. At 48 h RAW264.7 pre-osteoclasts were seeded in the lower chamber at a density of 5,000 cells/cm^2 in Dulbecco's Modified Eagle Medium (DMEM) with 10% FBS and 1% penicillin/streptomycin. After 24 h, osteoclasts were induced to undergo osteoclastogenesis with the addition of 120 ng/ml RANKL to the medium. Osteoclasts were fed with induction medium every 72 h using 100% medium replacement. Osteocytes were stimulated with polyethylene particles at 72 h and with oscillating fluid flow (1Pa, 1Hz) for 1 h every 24 h for 30 days. Conditioned medium was collected for analysis via ELISAs. The soluble factors, interleukin (IL)- 1β, and IL-6 were quantified. IL-1β enhances osteoclast differentiation while IL-6 indirectly promotes osteoclast differentiation by enhancing osteocyte-secreted RANKL (Fig. 7). At 30 days, bone wafers were removed from the LOC and assessed for TRAP activity and stained with toluidine blue for quantification of bone resorption (Fig. 8).

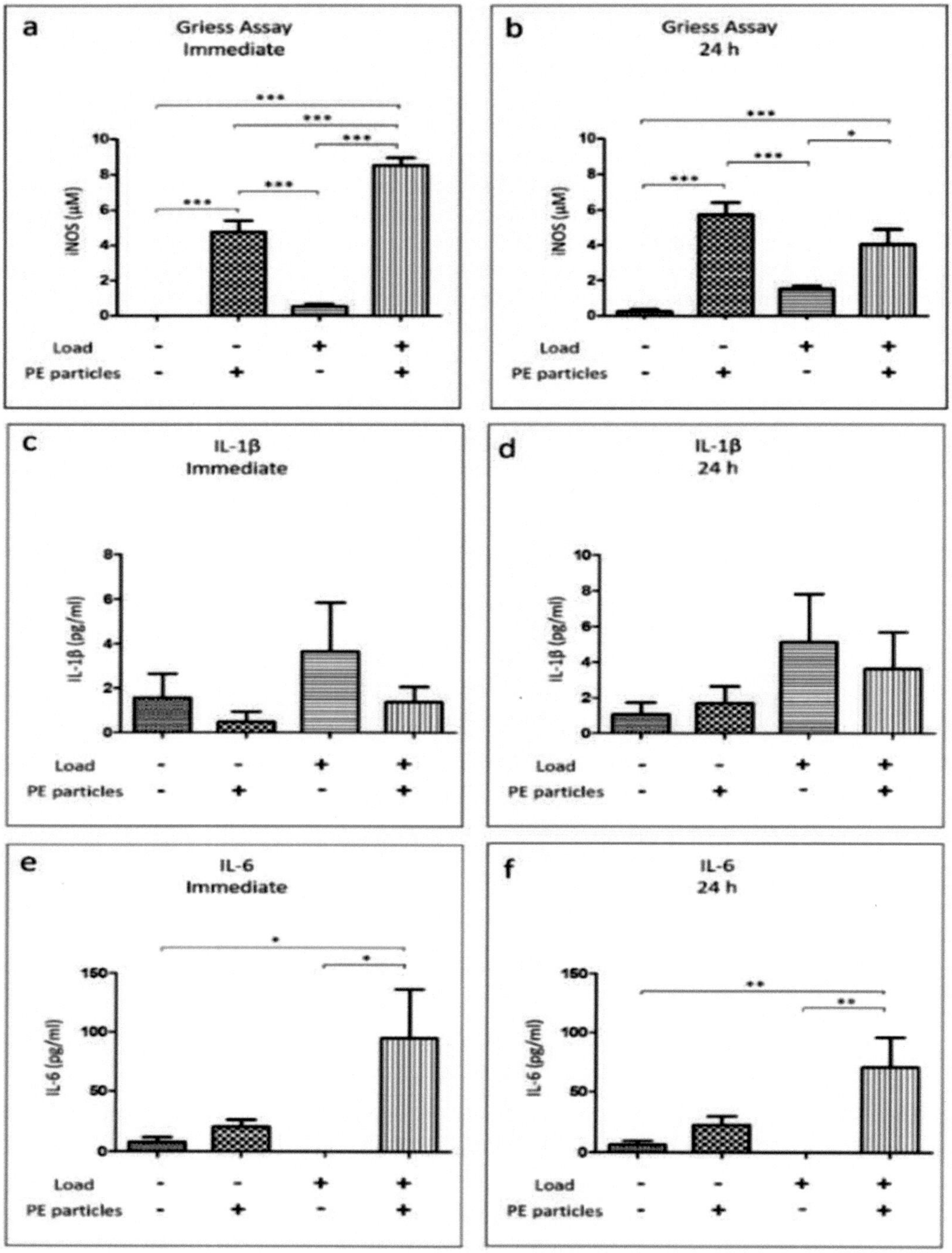

Fig. 7 Analysis results of osteocyte conditioned medium.

Results

To date, osteocyte conditioned medium assays showed the effects of combined loading and debris upon immediate collection or following a 24 h incubation period. Initial results suggested that in the LOC the combined effects on inflammation (osteocytic iNOS production) were largest immediately following collection. These effects decreased after 24 h but were not significantly different from the effects of particulate debris alone. Results from IL-1β quantification suggested that this cytokine production was not significantly different under the various combinations of load and particulate debris at the

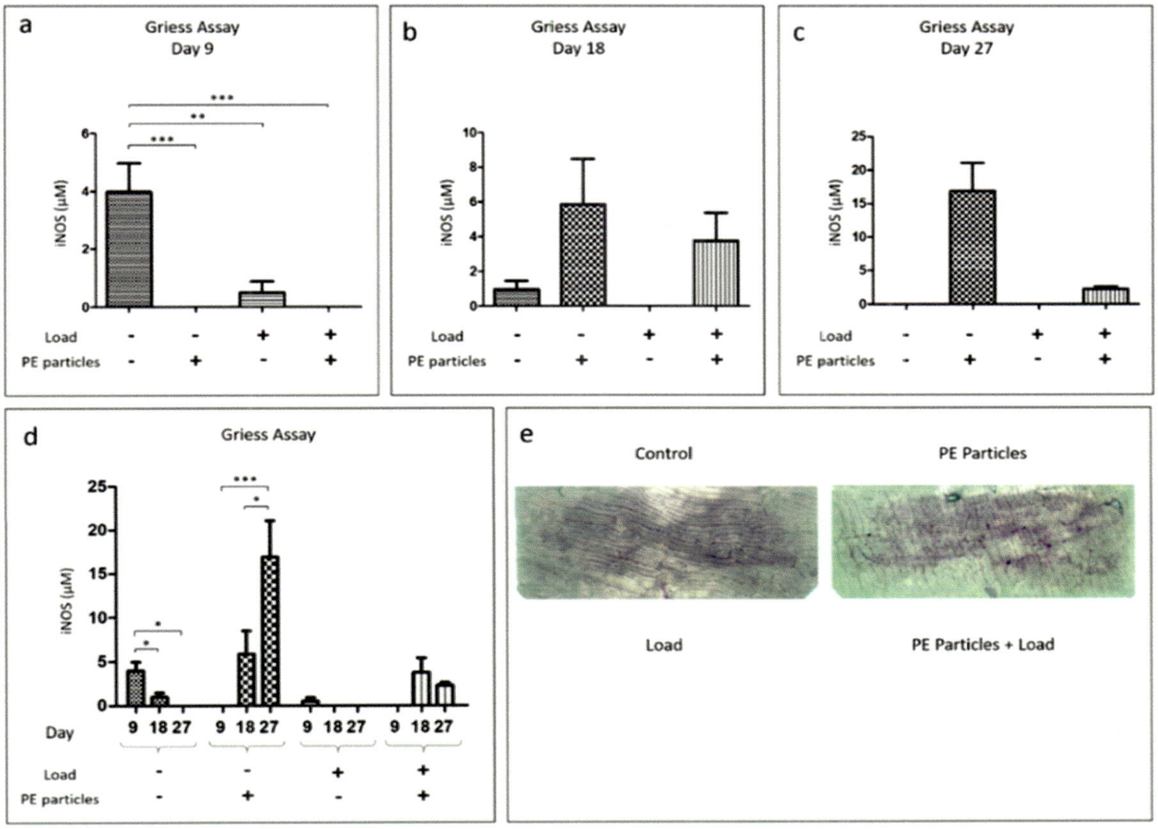

Fig. 8 Osteoclast iNOS production and bone resorption.

various collection times, while IL-6, which more indirectly promotes osteoclast activity by acting directly on RANKL, was significantly increased in combined cultures over baseline. To date, osteoclast conditioned medium assays showed that when osteocytes were not exposed to polyethylene particles, iNOS levels in the osteoclast chamber peaked at day 9 and then dropped significantly. But when osteocytes were exposed to polyethylene particles, OS levels in the osteoclast chamber were not detectable at day 9 but then increased significantly by day 27. Wafer resorption analysis has not been completed. While we are still working to define these results in the context of current knowledge, similar trends are emerging for the collection periods. However, the biomarker activity will help us to better understand the added benefit of the indirect measurements in the absence of the resorption data. This is also being completed to better vet the LOC and loading system for a wide range of research uses where it will be necessary to first establish true fidelity of the system and approach.

Conclusion

Although osteocytes make up the vast majority of bone cells and are known to play an essential role in regulating osteoclast and osteoblast activity, surprisingly little is known regarding their role in PPOL. In contrast, the biological response of macrophages to implant debris has been well established. Significant clinical evidence has demonstrated that wear debris from implants stimulate macrophages and trigger the release of pro-inflammatory cytokines, such as tumor necrosis factor (TNF)-α, interleukin (IL)-1, and IL-6 [7–8]. This inflamed environment ultimately leads to enhanced osteoclast activity and subsequent bone loss. While it is widely accepted that implant debris is a major cause of PPOL, some studies have suggested that mechanical stimulation may provide an additional mechanism leading to osteolysis that is independent of wear debris [9–10]. Researchers have hypothesized that micromotion at the bone-implant interface may result in a cyclic pressure-induced flow of joint fluid into the surrounding cancellous bone [11–12]. Further, some studies have reported that exposure to both wear particles and mechanical stimulation has a synergistic effect on pro-inflammatory cytokine expression in macrophages [13–14].

Recent studies suggest that osteocytes may also contribute to the pro-inflammatory response and subsequent increase in osteoclast resorption [15–16]. Ormsby et al demonstrated that exposure of human primary osteocyte-like cells to either polyethylene or metal particles upregulated expression of pro-inflammatory cytokines TNF-α and IL-6 [16]. Exposure of MLO-Y4 osteocytes to polyethylene particles has been shown to induce an upregulation of receptor activator of nuclear factor kappa-B ligand (RANKL), which promotes osteoclast differentiation [15–16]. And TNFα and IL-6 were shown to be upregulated in RAW264.7 macrophages exposed to titanium particles. However, this upregulation was suppressed by two inhibitors of SPHKs (FTY720 and ABC294640) suggesting that release of TNFα and IL-6 depends on SPHK-2 activity [17]. In osteocytes, it remains to be seen if a similar mechanism regulates particle-induced expression of these cytokines.

Acknowledgments The author would like to acknowledge the contributions of all former undergraduate and graduate student researchers in the Bone Mechanobiology and Orthopaedic Biomechanics Laboratory at The University of Akron, especially Ms Sharon Truesdell. The author would also like to thank the NIH and NSF for past funding of this work.

References

1. Kremers, HM. Prevalence of total hip and knee replacement in the United States. *J Bone Joint Surg Am* 97(17):1386-97, 2015
2. Perez Alamino R, et al. Prevalence of periprosthetic osteolysis after total hip replacement in patients with rheumatic diseases. *Open Access Rheumatol: Res Rev* 4:57-62, 2012
3. Atkins GJ, et al. Role of polyethylene particles in peri-prosthetic osteolysis: a review. *World J Orthop* 2(10):93-101, 2011
4. Gallo J, et al. Osteolysis around total knee arthroplasty: a review of pathogenic mechanisms. *Acta Biomater* 9(9):8046-58, 2013
5. George EL, et al. Lab-on-a-chip platforms for quantification of multicellular interactions in bone remodeling. *Exp Cell Res* 365(1):106-18, 2018
6. Truesdell SL, Saunders MM. A low-cost 3D-printed pumping system for generating oscillatory flow in a multicellular lab-on-a-chip device. *Microfluid Nanofluid* 27(6) Article 42, 2023
7. Nich C, et al. Macrophages – key cells in the response to wear debris from joint replacements. *J Biomed Res A* 101(10):3033-45, 2013
8. Connors JP, et al. The role of the innate immune system in wear debris-induced osteolysis. *Bioengineering* 9(12):764, 2022
9. McArthur BA, et al. Mechanically induced periprosthetic osteolysis: a systematic review. *HSS J* 15(3):286-196, 2018
10. Goodman SB, Gallo J. Periprosthetic osteolysis: mechanisms, prevention and treatment. *J Clin Med* 8(12):2091, 2019
11. Alidousti H, et al. Do capsular pressure and implant motion interact to cause high pressure in the periprosthetic bone in total hip replacements? *J Biomech Eng* 133(12), 2011
12. Camine VM, et al. A model for micromotion-induced fluid flow at the bone-implant interface. *Comp Meth Biomech Bioeng* 17(Suppl1):52-3, 2014
13. Adams S, et al. Mechano-immunomodulation: mechanoresponsiveness changes in macrophage activity and polarization. *Ann Biomed Eng* 47:2213-31, 2019
14. Couto M, et al. The mechanisms underlying the biological response to wear debris in total joint. *Front Mater* 7, 2020
15. Holding CA, et al. The correlation of RANK, RANKL and TNF alpha expression with bone loss volume and polyethylene wear debris around hip implants. *Biomaterials* 27(30):5212-9, 2006
16. Ormsby RT, et al. Osteocytes respond to particles of clinically-relevant conventional and cross-linked polyethylene and metal alloys by up-regulation of resorptive and inflammatory pathways. *Acta Biomater* 87:296-306, 2019
17. Yang G, et al. SPHK-2 promotes the particle-induced inflammation of RAW264.7 by maintaining consistent expression of TNF-α and IL-6. *Inflammation* 41(4):1498-1507, 2018

Chapter 3
The Stress Relaxation Response of the Porcine Descending Aorta under Combined Normal and Torsional Loadings

Luc Nguyen, Abdelrahman Youssef, Calvin Nguyen, Kirtan Patel, Jack Luce, Benjamin Tijerina, and Chandler C. Benjamin

Abstract Understanding the mechanical properties of aortic tissue is crucial for improving future medical interventions related to cardiovascular diseases. Our group has previously investigated the uni-axial, bi-axial, and shear responses of the porcine thoracic aorta. In this study, we explore the stress relaxation of the porcine aorta under torsional shearing while maintaining a constant normal compressive load. Circular samples, measuring 0.5 inches in diameter, were extracted from the descending section of porcine thoracic aortas. Stress relaxation experiments were conducted at a constant shear strain ranging from 10 to 50% while maintaining a compressive strain ranging from 5 to 25%. We observed that the stress relaxation behavior differs between the normal and torsional shear directions. Results suggest that neither the normal or shear stress relaxation behavior of the porcine aorta is significantly influenced by the degree of compressive strain or shear strain applied.

Keywords Stress relaxation · Creep · Aorta · Viscoelastic · Rheology

Introduction

A better framework for mechanical analysis of aortic tissue is needed to improve medical diagnosis and intervention. Current pathology for aortic diseases (such as aortic aneurysms, dissections, and stenosis) revolve around the determination of other diseases or genes that increase risk [16, 12, 20, 3]. However, the weakening of the tissue seen in aortic aneurysms and the rupture initiation seen in aortic dissections are clear signs that the pathology of aortic diseases needs to address mechanics.

Luc Nguyen · Abdelrahman Youssef · Calvin Nguyen · Chandler C. Benjamin
Department of Mechanical Engineering, Texas A&M University
email: lucnguyen314@tamu.edu; abdelyoussef@tamu.edu; calvinnguyennn@tamu.edu; ccbenjamin@tamu.edu

Kirtan Patel
Department of Biomedical Sciences, Texas A&M University
email: kpatel23@tamu.edu

Jack Luce
Department of Biology, Texas A&M University
email: jluce02@tamu.edu

Benjamin Tijerina
Department of Allied Health, Texas A&M University
email: btijerina2@tamu.edu

Chandler C. Benjamin
School of Engineering Medicine, Texas A&M University
Texas A&M University, College Station, TX, 77843
email: ccbenjamin@tamu.edu

Karen Kasza et al. (eds.), *Mechanics of Biological Systems and Materials and the Mechanics of Composite, Hybrid & Multifunctional Materials, Vol. 3*, Conference Proceedings of the Society for Experimental Mechanics Series, https://doi.org/10.13052/9788743808299-3

Observed mechanical properties of the aorta

In 1881, Charles Roy examined the elastic and viscous properties of animal arterial tissue [28]. His experiments demonstrated the first findings of viscoelastic material properties in arterial tissue. In 1969, Patel et al. noticed anisotropy in canine arteries in their inflation experiments. They measured differing mechanical responses in the circumferential and axial directions. They concluded that the thoracic aorta, abdominal aorta, and common carotid artery in canines can all be approximated as an orthotropic cylindrical tube [22]. From such previous work, we see the importance that anisotropy and viscoelasticity have in this material which must be considered in appropriate constitutive modeling for the aorta.

There have been many tests on the porcine aorta focusing on the uni-axial and bi-axial properties [18, 24, 21, 4, 7]. However, it has been observed that the shear strains induced by inflation are not negligible [22]. Coupling of wall shear stress and circumferential shear stress has also been observed in arterial tissue [29]. Torsional shear testing in our lab [26] has shown a nonlinear response between the experienced torsional shear stress and the angle of twist. We have additionally noticed a nonlinear coupling between the experienced torsional shear values and combined normal loadings [19]. In these previous tests in our lab, preconditioning was used to examine the elastic properties of the porcine aorta in its relaxed state. However, the time-dependent viscoelastic properties of the porcine aorta are crucial to proper modeling [17, 25].

Stress relaxation and creep in soft tissue

It has been shown that most basic viscoelastic models, such as Maxwell or Kelvin-Voight, fail to capture the response of biological tissue [10]. In their uniaxial tensile tests, Chen et al. note the stress relaxation measured on strips of porcine aortas at different angles [4]. Samples cut at the same angle with respect to the circumferential axis showed similar relaxation behavior, whereas samples at varying angles showed different stress relaxation behavior. The smallest relaxation occurred in the directions of the tissue fibers. As such, we expect anisotropy to be present even within the viscoelastic properties of the constitutive modeling for the aorta. In their inflation tests on canine carotid arteries, Zatzman et al. postulate the micro-constituents of the tissue to be responsible for the inelastic responses in their data [33]. Anssari-Benam et al. related their experimental stress relaxation data of the aortic valve to a kinematic framework of the fibers [1]. Their framework offers an explanation of the microstructural mechanics resulting in both the relaxation and creep phenomena. Bagshaw and Attinger compare the experienced stress relaxation in various canine arteries to the collagen and smooth muscle content [2]. They found a strong linear correlation between the experienced relaxation and the sum of collagen and smooth muscles content. We see that the phenomena of stress relaxation and creep are strongly tied to the microstructure of the artery. There is, however, a lack of experimental viscoelastic data of arteries under combined loadings.

Efforts to model aortic tissue

There has been much effort to model aortic tissue, yet each proposed model fails to capture all aspects of the aorta's mechanical properties let alone predict previously unobserved phenomena. Rajagopal and Rajagopal [25] discuss the complexities with modeling such a material. They note that the physical composition of the aorta is the source of many of the complexities as well as it's varying geometry and anisotropy.

Craiem et al. developed a fractional-order relaxation function [6]. They were able to fit uniaxial tensile stress relaxation data using quasilinear viscoelastic theory developed by Fung [11]. Fung et al. discuss the application of an exponential and polynomial strain energy function in fitting experimental data [9]. A given function with enough coefficients can be fit to most experimental data, and Fung et al. do note considerable variation between constants for different sets of data of the same experimental protocol. Tanaka and Fung utilized Fung's form for the reduced relaxation function [11] for canine arteries in cyclic uniaxial extension [31]. They note, however, that "we have no right to expect the quasi-linear relationship to be valid for full range of the inelastic phenomenon in large deformation." This is crucial as the aorta is expected to undergo large deformations.

Cheung and Hsiao developed a nonlinear, viscoelastic, anisotropic constitutive equation [5]. Some of the complexities of the composition of arteries are considered by modeling the material body as a system of micro-elements in equilibrium in their unloaded state. Their integral form of the relaxation function captures the inflation response of the canine carotid artery. Cheung and Hsaio note that only quasi-static conditions are considered and that the inertia term in the equation of motion should be taken into account with future work. Young et al. develop a nonlinear, viscoelastic, orthotropic model [32] where they express the stresses in canine aortas using integrals of four or ten relaxation functions and histories of circumferential and longitudinal strains. A motion comprising of internal inflation and longitudinal extension is examined to ensure no shear forces are present. While this allows for the use of their orthotropic constitutive relation, it greatly simplifies the stresses experienced by arteries. Holzapfel et al. proposed a rate-type constitutive relation based on the additive split of the second Piola–Kirchhoff stress into elastic and viscoelastic components [14]. They developed a finite element formulation modeling the arteries as fiber-reinforced composites. Their model successfully describes the hysteresis that occurs with cyclic loading that is insensitive to strain rate over several decades. Pena et al. used a thermodynamic framework in their constitutive relation to account for the "pseudo-elastic" behavior that vascular tissues demonstrate after preconditioning [23]. They used a simple neo-Hookean term to model the isotropic response and Holzapfel et al.'s exponential terms [15] to model the anisotropic terms. Pena et al note the following limitations to their constitutive formulation: a need for a large sample size for determination of material constants, omission of permanent set on stress removal due to loading-unloading tests, and the evolution equations of the viscoelastic internal variables are linear. This means that it is not possible to associate such internal variables with any internal micro-structural damage or thermodynamic processes. Haslach proposed a model using a system of evolution differential equations [13]. His model is able to predict both creep and stress relaxation under uniaxial extension. His model operates under the assumption that a unique long-term manifold exists for both creep and relaxation. Zerpa et al. [34] apply fractional viscoelastic models using "high-order spring-pot" to model the ovine ascending aortic wall. They note that the higher-order spring-pots leads to improvement of fitment to experimental data over traditional spring-pots. Zerpa et al. note that future studies should investigate histological interpretations of these higher-order spring-pots.

In general, it is clear that much work is to be done in the way of modeling aortic tissue. More experimental data is needed to further understand the types of mechanical behavior we expect to see from such a complex material. This proceeding focuses on the experimental data set obtained from loading porcine descending aorta in a combination of normal compression and torsional shear.

Experimental Methods

Porcine aortas were obtained from the Texas A&M Department of Animal Sciences. For this experiment, we only used samples from the descending thoracic aorta between the 3rd and the 7th branching arteries. These sections of aorta were longitudinally cut using a surgical scalpel, then circular samples were extracted using a half-inch circular punch. Samples that were visually non-uniform in thickness were discarded. Uniform circular samples were cleaned of excess tissue, placed in a vial filled with Phosphate Buffer Solution (PBS), then stored in a freezer maintained at -20° C. The samples remained frozen for at least 24 hours before testing. It has been previously determined that this method of preservation has no significant impact on the mechanical properties of the porcine aorta [21].

Experimentation was performed on an Anton Parr MCR302 rheometer. The rheometer was set up in a parallel plate configuration. We used a tissue bath for the fixed bottom plate and a 12.5mm diameter top plate. 120-grit sandpaper was attached to both the top and bottom plates to prevent slipping of the test specimen. Before each experiment, the sample was thawed by placing the frozen vial in room temperature water for at least 20 minutes. Measurements of the diameter and thickness of each circular sample were taken using Vernier calipers. Each sample was placed in the rheometer with the intimal side facing up and then visually centered underneath the top plate. The top plate was moved downward until contact with the sample was confirmed. The tissue bath was filled with PBS, and a hood was closed to maintain a temperature of 25° C.

Relaxation testing comprised of an axial normal compression followed by torsional shearing. The initial normal compression of either 5%, 10%, 15%, 20%, or 25% compressive strain was applied to the specimen over the span of 10 seconds. Torsional shearing of either 10%, 20%, 30%, 40%, or 50% was then applied over the span of 1 second. The constant normal strain and torsional shear strain was then maintained for 1000s. Ten samples were experimented on at each combination of normal and torsional shear strain for a total of 250 samples.

Results

Due to microstructural reconfigurations within the tissue, we expect the experienced stress to asymptotically relax to a non-zero value. The rheometer measured the experienced normal stress and shear stress from each sample. For each combination of normal compressive strain and shear strain, the data from the ten samples was averaged together for further analysis.

We normalized the data by plotting the measured stress divided by the maximum experienced stress on the ordinate. From initial observation of our results, we cannot confidently state that varying the normal or shear strain affects the relaxation behavior. For further analysis, we will examine the quasi-linear aspects of the results by averaging the data at varying compressive strains together. The results of this are shown in Figure 1. We notice in Figure 1, that the varying shear strains do not appear to have a significant effect on the relaxation behavior. The plots of each shear strain from Figure 1 were averaged to produce Figure 2. From our results, we see that a quasi-linear model could capture the data reasonably well for the stress relaxation of the porcine aorta in both the normal and shear directions.

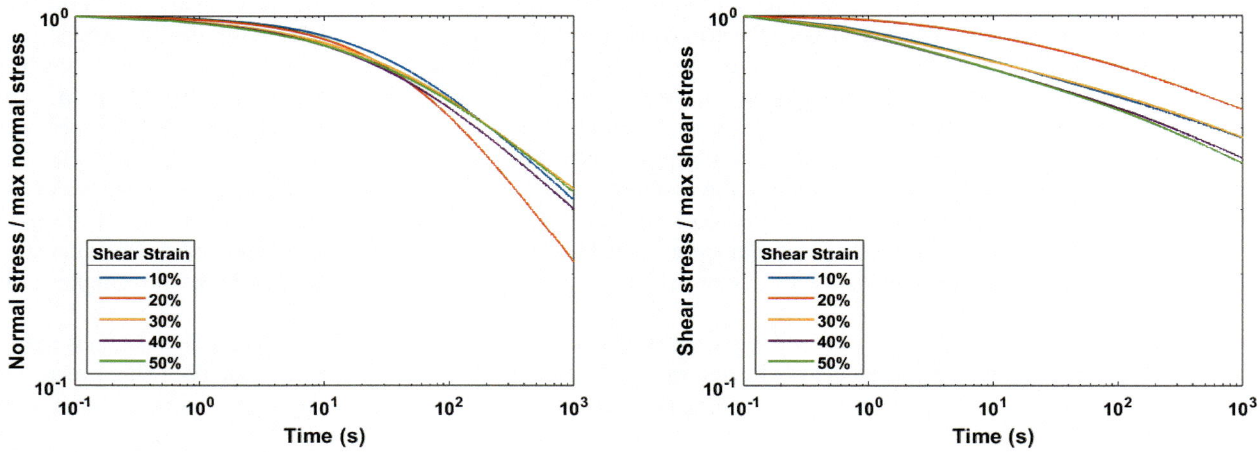

(a) Normalized normal stress over time

(b) Normalized shear stress over time

Fig. 1 Normalized (a) normal and (b) shear stress at varying shear strains after averaging all data at the same given normal compressive strain.

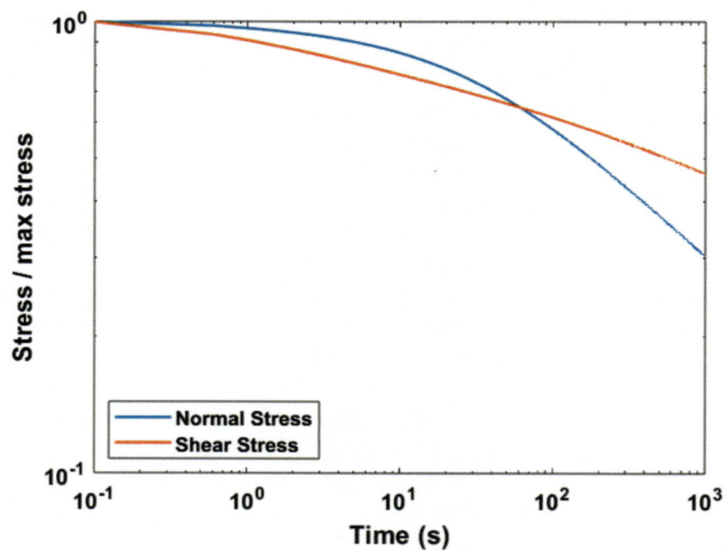

Fig. 2 Normalized normal and shear stress after averaging all data from Figure 1 at the same given shear strain.

Analysis

The normal relaxation modulus and the shear relaxation modulus are given by,

$$E(t) = \frac{\sigma(t)}{\epsilon_0} \tag{1}$$

$$G(t) = \frac{\tau(t)}{\gamma_0}, \tag{2}$$

where ϵ_0 and τ_0 denote the applied normal strain and shear strain, respectively. We normalized the data by

$$f(t;\vec{p}) = \frac{\sigma(t)}{\sigma_0} \quad , \quad g(t;\vec{p}) = \frac{\tau(t)}{\tau_0}. \tag{3}$$

Here, *f(t)* represents the normalized normal stress and *g(t)* represents the normalized shear stress. σ_0 denotes the initial normal stress at (the normal stress at $t = 0^+$), and τ_0 denotes the initial shear stress. We will model the normalized experimental data using a least-squares regression analysis. We will choose to model the data with

$$f(t;\vec{p}) = A + B\exp\left(\frac{-t}{\lambda}\right)\left(\frac{t}{\lambda}\right)^n. \tag{4}$$

Equation 4 is a modified version of the Weibull distribution [30]. Other variations of the Weibull distribution have been used to measure the creep and stress relaxation in polymeric materials [8]. In such models, we liken the viscoelastic, microstructural elements to a population of time-dependent mechanical latches, each with their own time to failure. The parameters n and λ are the shape and characteristic time parameters, respectively. A and B are coefficients. We seek to model the normalized data given in figure 2. We chose the modified Weibull distribution (equation 4) based on the shape of the relaxation data seen in figure 2. We expect an equation of exponential form to fit the data reasonably well. For an initial study, we are seeking to find a nonlinear equation that can be curve fit to the data with a reasonably good coefficient of determination (R-squared value).

The vector $\vec{p}$ denotes the material parameters which, in equation 4, are

$$\vec{p} = \begin{bmatrix} A \\ B \\ \lambda \\ n \end{bmatrix}.$$

We will model both the normal and shear relaxation data using equation 4. A nonlinear least squares regression produces the best-fit parameters for $\vec{p}$ given in Table 1. The experimental data, the modeled equation and the R-squared values are shown in Figure 3.

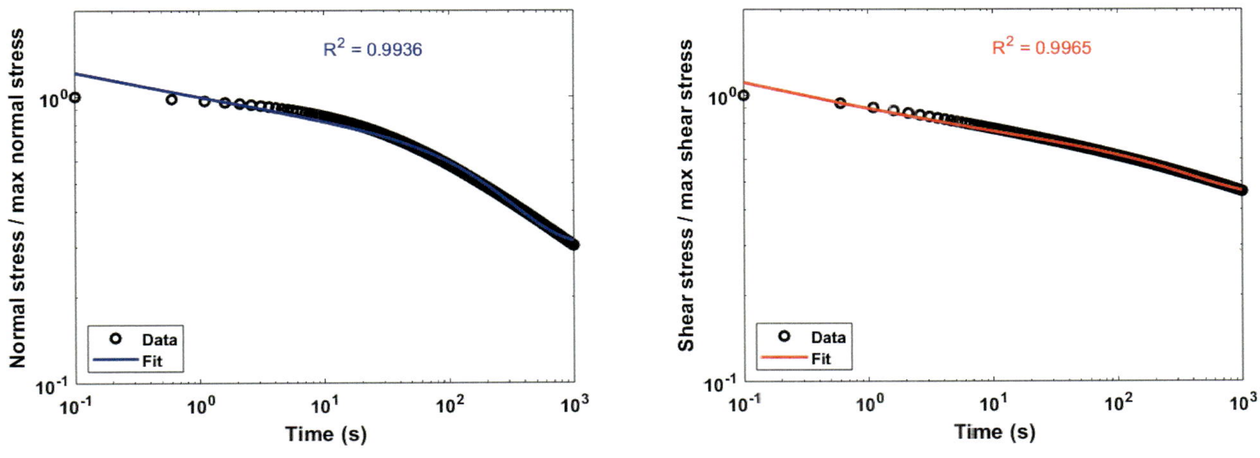

(a) Normalized normal stress over time

(b) Normalized shear stress over time

Fig. 3 The data and fit using a least-squares regression analysis of the experimental data and equation 4 for the normalized (a) normal and (b) shear stress.

We notice the characteristic time parameter, λ, differs between the normal and shear directions. We also note that, even with a nonlinear least-squares regression analysis, the model fails to perfectly capture the data at t = 0.1s. Because Figure 3 is on a log-log scale, the difference between the model and data becomes exaggerated compared to a linear scale. Next, we examine the results using linear-log axes to determine the effects of varying the magnitude of compressive strain and shear strain on the relaxation response.

Table 1 Solved material parameter values

Direction	A	B	$\lambda(s)$	n
Normal	0.305	0.356	293	−0.117
Shear	0.448	0.162	488	−0.166

Figure 4 shows all raw data plotted on a semi-log scale. From the data, we do not observe any clear effect of the shear strain amplitude on the stress relaxation response in either the normal or shear directions. The characteristic time generally appears to remain the same regardless of the input shear strain chosen. Considering the effect of the compressive strain amplitude, we observe a generally positive correlation between the nominal stress values and the given compressive strain. However, after normalization (Figure 1), we cannot state a clear significant effect that varying compressive strain has on the relaxation behavior.

Discussion

Implications of Experimental Data

From our normalized data, we notice there is not a significant difference in stress relaxation behavior due to varying the input normal strain nor the input shear strain. With ten samples at each combination of control variables, we believe this experiment sufficiently demonstrates that the viscoelastic response of the porcine aorta can be approximated using a quasi-linear viscoelastic model. We also note the difference in the shape of the trend lines between the normal and shear stress response. This demonstrates the anisotropy of the material.

In our analysis, we see that both data sets in the normal and shear directions can be fitted using a model of the form of equation 4. At the time of this publishing, we have not proved uniqueness to this model. In fact, we do not expect uniqueness from this model. Further study is required for the viscoelastic constitutive modeling of the porcine aorta. The purpose of this paper is to provide an experimental data set in which future modeling can be validated with.

From this data set, we see the magnitude of the time-dependence of the material in combined normal and shear loading. Within the time frame of the experiment (1000 seconds), we can further examine the viscoelastic behavior of aortic tissue. From Figure 3, we can see that neither the normal or the shear stress plateau to a finite stress value. From longer experiments, we can determine whether the material behaves as a viscoelastic solid or viscoelastic fluid.

Novelty of our experiment

The novelty of this experiment comes from both the loading configuration and the combination of normal and shear forces onto the test samples. We did not test samples of porcine aortas in their natural, exhumed configuration. In fact, we do not expect normal physiological loadings in the configuration in which we are testing. Additionally, with the way the samples are prepared, we expect residual stresses to exist within the body. Thus, the nominal values of our experiments will not correlate to the stress values we expect to see in vivo. However, the purpose of this experiment is to gather an understanding of the viscoelastic response the porcine aorta shows under combined loading. Our methods allow us to take multiple samples from each aorta. This allows for a wider range of loading conditions and a larger number of samples at each combination. As a biological material, we expect each sample of porcine aorta to show large variance, and testing at least 10 samples at each loading condition is necessary for characterizing the material. While statistical accuracy is expected to be low for a biological material, we can minimize this by taking many samples out of each aorta retrieved.

Implication on future modeling

We discovered a clear anisotropic relaxation response within the porcine aorta. From the relaxation experiment, we determine if the material behaves as a viscoelastic solid or fluid. This will impact the choice between a Gibbs or Helmholtz free energy [27]. From our findings, we determine that future modeling of the porcine aorta must include anisotropic relaxation behavior. The rate of dissipation will need to vary in the normal and shear directions.

Questions our data pose

From figure 4, we see that the averaged data of the different compressive strains do not behave entirely as expected. In some cases, such as the data set run at 20% shear strain, we see that 10% compressive strain experiences less normal stress than

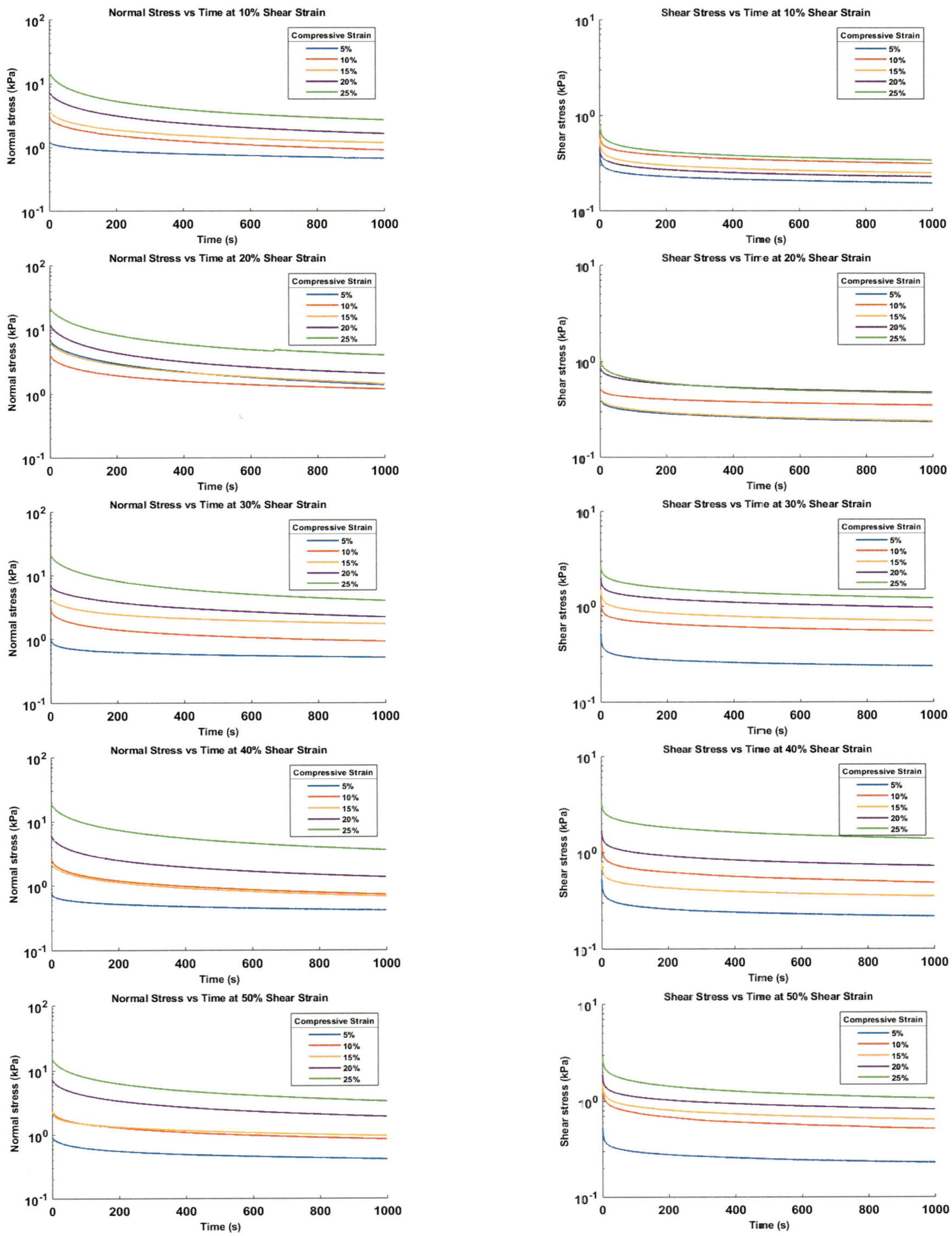

(a) Normal stress over time

(b) shear stress over time

Fig. 4 Semilog plots of (a) normal stress and (b) shear stress as a function time at varying shear strain and varying compressive strain.

5% compressive strain. If we examine the final stress (the stress value at the end of the test) in relation to the changes in compressive strain, shown in figure 5, we do not see any definitive correlation. We expect there to be a high variance within the different biological samples, so the question arises as to whether this peculiarity in the data is a result of the large spread in experimental data or if there is something within the microstructure of the porcine aorta that leads to such a phenomenon.

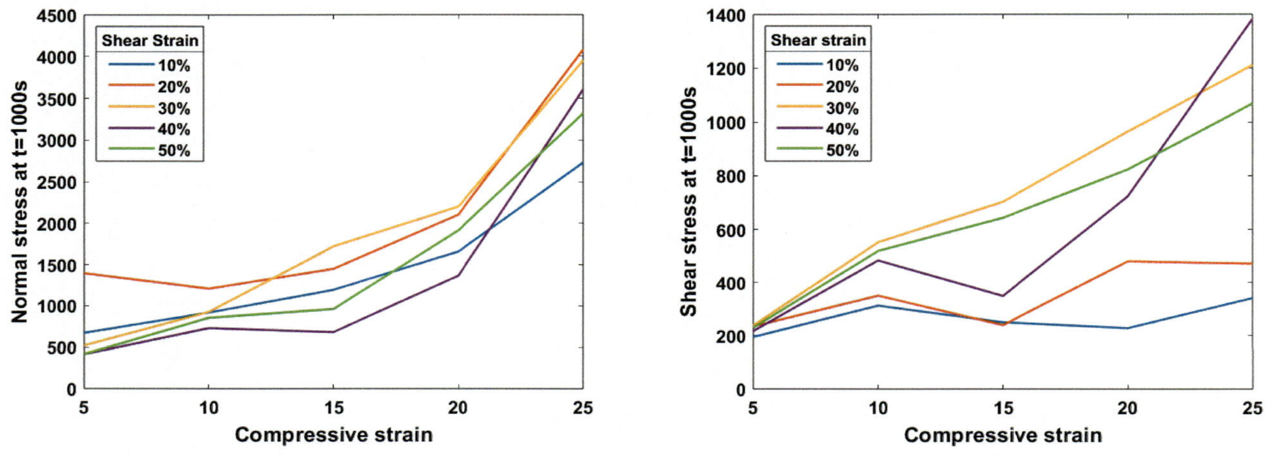

(a) Final normal stress vs applied compressive strain

(b) Final shear stress vs applied compressive strain

Fig. 5 The (a) normal and (b) shear stress at time t = 1000s as a function of the applied compressive strain. The varying shear strains are plotted on the same figures.

Conclusions

We noticed the relaxation behavior is different in the normal and shear directions. The amplitude of compressive strain appears to have a larger impact on the quantitative experienced stress compared to the amplitude of the applied shear strain. There appears to be some nonlinear correlation between the normal and shear stress responses of the porcine descending aorta. Further investigation is required to better characterize this correlation.

Future work

Future work for this project will include a creep experiment to examine if the creep behavior can be predicted using this relaxation data. This creep experiment will be conducted using similar testing conditions as those found in this paper. We expect the porcine aorta to behave as a nonlinear material, so the creep behavior should not be able to be directly predicted from the relaxation data. Creep experimental data will be needed to confirm such hypotheses. An analysis on the kinematics and force balances will also be done for the problem given by this experimental set up. With the creep and relaxation data and the mechanics analysis, further improvements on constitutive modeling can be made.

The purpose of this study is to provide a robust set of experimental data on the relaxation response of the porcine aorta. As such, testing time was limited to 1000 seconds per sample. In the future, we seek to test samples until the value of the stress plateau in both the normal and shear directions. This will help to determine with certainty if the porcine aorta acts as a viscoelastic solid or viscoelastic fluid. We also seek to perform histology analysis on samples used in these relaxation experiments. This will further examine how the relaxation response of the porcine aorta relates to the permanent damage to the tissue.

References

1. Afshin Anssari-Benam, Hazel RC Screen, and Andrea Bucchi. Insights into the micromechanics of stress-relaxation and creep behaviours in the aortic valve. Journal of the mechanical behavior of biomedical materials, 93:230–245, 2019.
2. RJ Bagshaw and Françoise ML Attinger. Longitudinal stress relaxation in the canine aorta. Experientia, 30(9):1046–1047, 1974.

3. Thierry Carrel, Thoralf M Sundt, Yskert von Kodolitsch, and Martin Czerny. Acute aortic dissection. The Lancet, 401(10378):773–788, 2023.
4. Qiang Chen, Yan Wang, and Zhi-Yong Li. Re-examination of the mechanical anisotropy of porcine thoracic aorta by uniaxial tensile tests. BioMedical Engineering OnLine, 15(2):493–506, 2016.
5. JB Cheung and CC Hsiao. Nonlinear anisotropic viscoelastic stresses in blood vessels. Journal of biomechanics, 5(6):607–619, 1972.
6. Damian Craiem, Francisco J Rojo, Jose Miguel Atienza, Ricardo L Armentano, and Gustavo V Guinea. Fractional-order viscoelasticity applied to describe uniaxial stress relaxation of human arteries. Physics in Medicine & Biology, 53(17):4543, 2008.
7. Valèrie Deplano, Mourad Boufi, Olivier Boiron, Carine Guivier-Curien, Yves Alimi, and Eric Bertrand. Biaxial tensile tests of the porcine ascending aorta. Journal of biomechanics, 49(10):2031–2037, 2016.
8. Kevin S Fancey. A mechanical model for creep, recovery and stress relaxation in polymeric materials. Journal of materials science, 40(18):4827–4831, 2005.
9. YC Fung, K Fronek, and P Patitucci. Pseudoelasticity of arteries and the choice of its mathematical expression. American Journal of Physiology-Heart and Circulatory Physiology, 237(5):H620–H631, 1979.
10. Yuan-cheng Fung. Biomechanics: mechanical properties of living tissues. Springer Science & Business Media, 2013.
11. Yuan-Cheng B Fung. Stress-strain-history relations of soft tissues in simple elongation. Biomechanics its foundations and objectives, pages 181–208, 1972.
12. Jonathan Golledge and Paul E Norman. Atherosclerosis and abdominal aortic aneurysm: cause, response, or common risk factors?, 2010.
13. Henry W Haslach Jr. Nonlinear viscoelastic, thermodynamically consistent, models for biological soft tissue. Biomechanics and Modeling in Mechanobiology, 3(3):172–189, 2005.
14. Gerhard A Holzapfel, T Christian Gasser, and Michael Stadler. A structural model for the viscoelastic behavior of arterial walls: continuum formulation and finite element analysis. European Journal of Mechanics-A/Solids, 21(3):441–463, 2002.
15. Gerhard A Holzapfel, Thomas C Gasser, and Ray W Ogden. A new constitutive framework for arterial wall mechanics and a comparative study of material models. Journal of elasticity and the physical science of solids, 61:1–48, 2000.
16. Deepali Jain, Harry C Dietz, Gretchen L Oswald, Joseph J Maleszewski, and Marc K Halushka. Causes and histopathology of ascending aortic disease in children and young adults. Cardiovascular Pathology, 20(1):15–25, 2011.
17. Manoj Myneni and KR Rajagopal. Constitutive modeling of the mechanical response of arterial tissues. Applications in Engineering Science, 11:100111, 2022.
18. Manoj Myneni, Raghuveer Lalitha Sridhar, Kumbakonam R Rajagopal, and Chandler C Benjamin. Experimental investigation of the anisotropic mechanical response of the porcine thoracic aorta. Annals of Biomedical Engineering, 50(4):452–466, 2022.
19. Luc Nguyen, Manoj Myneni, Abdelrahman A Youssef, and Chandler C Benjamin. Effect of normal stress on the torsional shear response of the porcine descending aorta. In Society for Experimental Mechanics Annual Conference and Exposition, pages 21–24. Springer, 2022.
20. Christoph A Nienaber, Rachel E Clough, Natzi Sakalihasan, Toru Suzuki, Richard Gibbs, Firas Mussa, Michael P Jenkins, Matt M Thompson, Arturo Evangelista, James SM Yeh, et al. Aortic dissection. Nature reviews Disease primers, 2(1):1–18, 2016.
21. Siobhan A O'Leary, Barry J Doyle, and Tim M McGloughlin. The impact of long term freezing on the mechanical properties of porcine aortic tissue. Journal of the mechanical behavior of biomedical materials, 37:165–173, 2014.
22. Dali J Patel, Donald L Fry, and Joseph S Janicki. The elastic symmetry of arterial segments in dogs. Circulation research, 24(1):1–8, 1969.
23. E Peña, V Alastruè, A Laborda, MA Martìnez, and M Doblarè. A constitutive formulation of vascular tissue mechanics including viscoelasticity and softening behaviour. Journal of biomechanics, 43(5):984–989, 2010.
24. Juan A Peña, Miguel A Martìnez, and Estefanìa Peña. Layer-specific residual deformations and uniaxial and biaxial mechanical properties of thoracic porcine aorta. Journal of the Mechanical Behavior of Biomedical Materials, 50:55–69, 2015.
25. Kumbakonam R Rajagopal and Keshava Rajagopal. Modeling of the aorta: complexities and inadequacies. Aorta, 8(04):091–097, 2020.
26. Akshay Rao, Manoj Myneni, CC Benjamin, and KR Rajagopal. High amplitude torsional shear of porcine thoracic aorta. In Mechanics of Biological Systems and Materials & Micro-and Nanomechanics, Volume 4, pages 37–40. Springer, 2020.
27. IJ Rao and KR Rajagopal. A thermodynamic framework for the study of crystallization in polymers. Zeitschrift f`ur angewandte Mathematik und Physik ZAMP, 53:365–406, 2002.
28. Charles S Roy. The elastic properties of the arterial wall. The Journal of physiology, 3(2):125, 1881.
29. Milad Samaee, Mohammad Tafazzoli-Shadpour, and Hamed Alavi. Coupling of shear–circumferential stress pulses in-vestigation through stress phase angle in fsi models of stenotic artery using experimental data. Medical & Biological Engineering & Computing, 55(8):1147–1162, 2017.
30. Ammar M Sarhan and Mazen Zaindin. Modified weibull distribution. APPS. Applied Sciences, 11:123–136, 2009.
31. Toshiyuki T Tanaka and Yuan-Cheng Fung. Elastic and inelastic properties of the canine aorta and their variation along the aortic tree. Journal of Biomechanics, 7(4):357–370, 1974.
32. John T Young, Ramesh N Vaishnav, and Dali J Patel. Nonlinear anisotropic viscoelastic properties of canine arterial segments. Journal of Biomechanics, 10(9):549–559, 1977.
33. M Zatzman, RW Stacy, J Randall, and A Eberstein. Time course of stress relaxation in isolated arterial segments. American Journal of Physiology-Legacy Content, 177(2):299–302, 1954.
34. JM Pèrez Zerpa, A Canelas, B Sensale, D Bia Santana, and Ricardo L Armentano. Modeling the arterial wall mechanics using a novel high-order viscoelastic fractional element. Applied Mathematical Modelling, 39(16):4767–4780, 2015.

Chapter 4
Investigating the Interphase in Hydroxyl-Terminated Polybutadiene (HTPB) Composites via Dynamic Mechanical Analysis and Atomic Force Microscopy

Jarred G. Tramell, Emily K. Hockey, Marcel M. Hatter, and Jesus O. Mares

Abstract State of the art explosive and propellant systems typically consist of energetic particles and a binder system. Specialized systems may contain additional components to meet application requirements, but in general these systems can be classified as highly filled polymer composite materials. The most extensively used binder systems across the energetics community are hydroxyl-terminated polybutadienes (HTPB). The HTPB binder functions to hold the particulates together into a consolidated structural media. It is well established that HTPB binders can significantly influence mechanical behavior, but much of the understanding of the fundamental mechanisms at play is lacking. Specifically, there exists a knowledge gap for the role of the interphase between the particles and the binder (i.e. the transition zone between pure binder and a particulate).

Dynamic mechanical analysis (DMA) is a technique that may shed light on the quality of the particle-binder interphase. Many polymer composites exhibit what appear to be two independent relaxation mechanisms via DMA experiments. There is general agreement that the first relaxation mechanism correlates to the glass transition temperature of the bulk polymer, however there is debate of the cause of the second relaxation mechanism. One prominent suggestion is that the secondary relaxation mechanism is attributed to the significant presence of an interphase region.

In this work we systematically investigate the effects of an effective interphase region on the presence of the loss factor peaks as exhibited through DMA experimentation. Specifically, we vary glass bead particle size in a HTPB-based composite and compare the loss factor to the measured effective volume of the interphase via atomic force microscopy (AFM). This approach may prove useful to understand the interphase mechanisms at play in energetic systems which will lead to the ability to create materials with tailored bulk mechanical properties and mitigate mechanical failures.

Keywords Hydroxyl-Terminated Polybutadiene · Filled Composite · Interphase · Dynamic Mechanical Analysis · Atomic Force Microscopy

Introduction

Hydroxyl terminated polybutadiene (HTPB) systems are some of the most common polymer matrices used for insensitive highly filled energetic systems such as polymer bonded explosives (PBX) and rocket propellants. HTPBs used in energetics typically consist of a crosslinked HTPB elastomer, plasticizer, and various stabilizers. The polymer matrix used in energetic systems is often described as the "binder" and primarily functions to (1) hold the particulates together into a consolidated structural media, (2) provide mechanical integrity to the system during loading scenarios and (3) reduce sensitivity of the energetic particles to mechanical stimuli.

In general, the interface between a particulate and the polymer matrix in filled composites creates an interphase region. This interphase region can be split into an inner layer and an outer layer. The inner layer contains polymer chains that are less

Jarred G. Tramell · Emily K. Hockey · Jesus O. Mares
Air Force Research Laboratory
e-mail: jarred.tramell.1@us.af.mil; emily.hockey.ctr@us.af.mil; jesus.mares.2@us.af.mil

Marcel M. Hatter
University of Dayton Research Institute, Eglin Air Force Base, FL 32542
e-mail: marcel.hatter.ctr@us.af.mil

Karen Kasza et al. (eds.), *Mechanics of Biological Systems and Materials and the Mechanics of Composite, Hybrid & Multifunctional Materials, Vol. 3*, Conference Proceedings of the Society for Experimental Mechanics Series, https://doi.org/10.13052/9788743808299-4

mobile than the bulk binder due to interaction with the particulate (i.e. tightly bound). The outer layer has higher mobility than the inner layer, but still less than that of the bulk binder (i.e. loosely bound). The reduction in molecular mobility and the sizes of the respective interphase regions depends on the polymer matrix used, and the particle's surface chemistry, volume fraction, particle size, particle size distribution, and quality of dispersion within the binder [1].

It is understood that the polymer matrix influences the overall mechanical behavior of the composite, but there is a lack of understanding of the how the interphase region between the HTPB binder and filler contribute to the overall mechanical behavior. Dynamic mechanical analysis (DMA) is a technique that may provide insight into the properties of the interphase. DMA provides information of the viscoelastic properties of a material, and specifically in this study tan δ (i.e. loss factor) is investigated. The tan δ represents the ratio between a material's loss modulus (energy dissipation) and storage modulus (energy storage). The tan δ provides insights into a material's ability to dampen vibrations or absorb impacts. A peak in a tan δ trace indicates that the material is exhibiting a significant energy loss. This energy loss is primarily attributed to increased molecular mobility within the material and is often associated with a specific relaxation process (e.g. the glass transition temperature). In filled HTPB energetic systems, many have reported the presence of two distinct tan δ peaks, which suggests the presence of two distinct relaxation processes [2–6].

The first tan δ peak at the coldest temperature is broadly accepted to describe the glass transition temperature (T_g) of the bulk HTPB binder. However, there are several explanations of the origin of the second-broader tan δ peak that occurs at warmer temperatures. The second tan δ peak is commonly attributed to a separate relaxation process solely within the binder. The most frequent explanation attributes the second peak to the hard and soft segments of the HTPB binder. Another explanation ascribes this to the presence of sol-gel content in the polymer network (i.e. non-crosslinked chains) [3–5]. Despite previously published explanations for the additional tan δ peak in filled HTPB systems, other filled composite systems often link this second peak to the presence of an interphase region [1]. Although not universally applicable to every filled system, the interphase relaxation mechanism seen in DMA data for many other systems warrants consideration as the underlying mechanism of the second tan δ peak in highly filled HTPB systems [1, 7–9].

Bashir discusses several alternative and complimentary approaches to DMA that other investigators have used to characterize the interphase region such as nuclear resonance spectroscopy (NMR), atomic force microscopy (AFM), Raman spectroscopy, and dielectric relaxation spectroscopy [1]. AFM is perhaps the most attractive option since the technique can measure the widths and modulus of the interphase regions and could be useful to correlate interphase characteristics to tan δ peak heights and widths from DMA data.

The selected approach of this work is to systematically change the particle size of filler and modify the particles' surface chemistry, both of which should change the volume fraction of interphase region [1]. Changes to the effective interphase volume are expected to create changes in DMA tan δ traces and AFM measurements. By pairing DMA and AFM measurements, it may be possible to determine if DMA can infer additional information on the interphase region in filled HTPB systems (e.g. the interphase size).

Materials

A typical HTPB binder system was selected as the polymer matrix. Table 1 summarizes the components of the HTPB system. Soda-lime silica glass spheres (glass beads) were selected as the filler instead of energetic particles to simplify testing logistics and estimation of total surface area of the solids in the system. The weight percent of glass beads was chosen to generate a system with 50 volume % of filler. The filler content for all mixes was fixed at 50 volume % and the sizes of the glass beads were varied.

Table 1 Filled HTPB composition

Ingredient	Function	Weight %
Poly bd® R45 HTLO	Prepolymer	18.41
Isodecyl pelargonate (IDP)	Plasticizer	6.138
Dibutyltin dilaurate (DBTDL)	Catalyst	0.04242
Ethyl 702	Antioxidant	0.1564
Isophorone diisocyanate (IPDI)	Curative	1.766
Soda-lime silica glass sphere	Filler	73.49

Soda-lime silica glass should have a high level of interaction with the binder due to the surface silanol groups at the beads' surface. The silanol groups can react with the IPDI and create a covalent bond between the filler and the polymer network.

To tune the surface chemistry of the glass beads, Sigmacote® was applied to the particles. Sigmacote® is a siliconizing reagent for glass and is intended to function in this system as a shielding agent to reduce particle-binder interaction effects and prevent particle-binder covalent bonding. Glass beads were purchased from Potter's Industries and Sigmacote® was purchased from Sigma-Aldrich. Mixes were generated with and without coated glass beads in this work. For the remainder of this study, mixes that utilize glass beads with Sigmacote® are denoted with the suffix "-SC".

Prior to mixing, glass beads were sieved to reduce the span of the particle size distribution. The grades of spheres selected in this work and the U.S. mesh sieves that the beads were collected between are shown in table 2. Table 2 also includes particle size data on the post processed glass beads. The particle size data confirms that the application of Sigmacote® did not significantly change the particle size distributions of the glass beads.

Table 2 Particle size data and surface area estimates for coated and uncoated glass beads

Filler	U.S. Mesh	Dv10 (μm)	Dv50 (μm)	Dv90 (μm)	Average Particle Size (μm)	Surface Area per mL of filler (cm^2/mL)
P4000	325,500	7.14	21.26	39.70	14.35	4181
P4000-SC	325,500	8.78	23.42	41.15	16.75	3582
P0040	120,170	59.66	69.25	81.08	66.5	902.2
P0040-SC	120,170	55.97	68.64	86.4	67.64	530.0
P0080	70,80	151.4	178.0	207.5	176.4	340.1
P0080-SC	70,80	149.1	174.6	205.6	173.3	346.2
P0170	40,45	340.1	378.5	412.5	375.3	159.8
P0170-SC	40,45	331.2	375.1	411.0	370.8	161.8
P0280	25,30	529.9	608.7	680.5	602.0	99.67
P0280-SC	25,30	523.5	606.1	695.8	602.5	99.59

Formulation components were weighed out and hand-mixed for two minutes at ambient temperature. After hand-mixing, the mixture was degassed in a vacuum chamber at ambient temperature for 10 minutes to remove air in the system. Larger sphere sizes showed minimal settling at the end of the degassing step and were gently mixed to resuspend particles to mitigate settling and reintroduction of air. Mixes were then poured into open face 3.2 x 12.5 x 60 mm molds and cured at 60 °C for 48 hours. Qualitatively, no settling was observed by eye in the cured samples.

Figure 1 shows a picture of cured samples. A color difference is observed between the coated and uncoated beads. At this time, it is unknown what caused the color change. Some potential explanations could be as follows: side reactions/molecular complexing between the Ethyl 702 or DBTDL and Sigmacote®, or refractive index differences between glass beads with and without covalent bonding to the binder. Currently, we have dismissed residual HCl from the coating reaction as the root cause, as coated beads were washed with deionized water and the rinse had a neutral pH. The underlying cause of the color may have an inconsequential effect on the results, since color changes can manifest at ppm levels, but it cannot be discounted when interpreting the results.

Methods

Dynamic mechanical analysis

DMA data was collected with a TA Instruments Discovery Hybrid Rheometer (DHR20) equipped with a rectangular torsional geometry. Samples were tested with a frequency sweep (0.01, 0.1, 1, and 10 Hz) from -110 to 100 °C with 1°C temperature steps and a 0.1% oscillatory strain. Samples were verified to be within the linear viscoelastic regime (LVR) prior to selecting the 0.1% oscillatory strain. A tensile axial force of 0.15 ± 0.1 N was maintained throughout the procedure to avoid skewing results from thermal expansion/contraction. Specimens were equilibrated for 45 seconds prior to data collection at each temperature step. Collected DMA data is parsed into tan δ vs temperature curves at each respective test frequency (0.01, 0.1, 1, and 10 Hz) for analysis.

DMA data processing

Since DMA traces for filled HTPB composites historically exhibit two overlapping tan δ peaks, such as the example shown in figure 2, it is necessary to decompose the traces into distinct peaks to properly quantify the peak heights, widths, and areas.

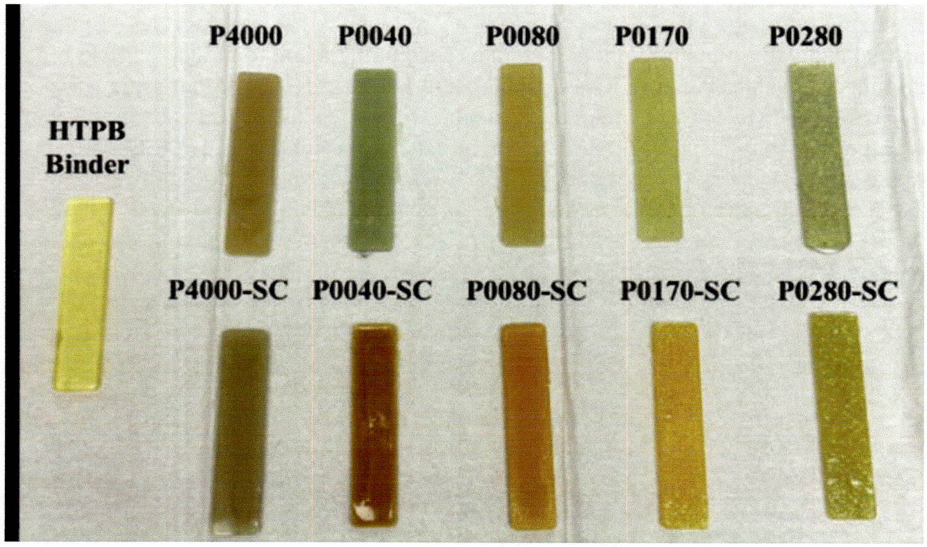

Fig. 1 Picture of cured DMA samples where the glass bead size increases from left to right and uncoated and coated beads are on the top and bottom of the image, respectively

Tsagaropoulos & Eisenberg first proposed modelling and decomposing tan δ traces with an exponential background and two exponentially modified Gaussian (EMG) distributions [9]. However, their approach has since been adapted By Cerri and Bohn towards HTPB systems which models tan δ traces with three EMG distributions to quantify the peak heights, widths, and areas [10]. To fit the tan δ traces to EMG curves, the data is baseline corrected then fitted to equation (1). Variables used in equation (1) are bulleted below with their associated units used. td_0 was set to zero for this work.

The effect of the glass bead particle size and coating on the parameters listed in equation (1) will be assessed. Any reduction in A_i is reported by Bohn *et. al.* to relate to the hindrance of molecular mobility in the system [4]. Additionally, the authors of this study speculate that larger w_i values correlate to a larger distribution of relaxation times and thus a larger interphase region.

$$\tan(\delta)_{BL} = td_0 + \sum_{i=1}^{N} \frac{A_i}{\tau_i} \bullet \frac{1}{2} \bullet exp\left[0.5 \bullet \left(\frac{w_i}{T_{0_i}}\right)^2 - \frac{T - T_{c_i}}{\tau_i}\right] \bullet \left\{1 - \mathrm{erf}\left[-\frac{1}{\sqrt{2}} \bullet \left(\frac{T - T_{c_i}}{w_i} - \frac{w_i}{\tau_i}\right)\right]\right\} \quad (1)$$

- T - measurement temperature (K)
- $\tan(\delta)_{BL}$ - value of tan δ after baseline correction as a function of T (-)
- A_i - EMG peak areas (K)
- w_i - half width at half height of the gaussian peak (K)
- T_{c_i} – temperature at Gaussian peak maximum (K)
- τ_i - relaxation parameter in exponential part of EMG (K)
- td_0 – offset in tan δ data (-)
- N – number of EMG fitting functions

The values for T_{c_i} at each Gaussian peak are also used to calculate the activation energy (E_a) of each relaxation process with the Arrhenius equation (shown in equation 2) where f is the applied test frequency, f_0 is a pre-exponential factor, R is the ideal gas constant, and $T(f)$ is the temperature at a given tan δ peak at each deformation frequency (T_{c_i} in this case). Calculating the E_a will provide further insight of the interphase since higher activation energies indicate stronger particle-filler interactions [11, 12].

$$f = f_0 \bullet exp\left(-\frac{E_a}{R * T(f)}\right) \quad (2)$$

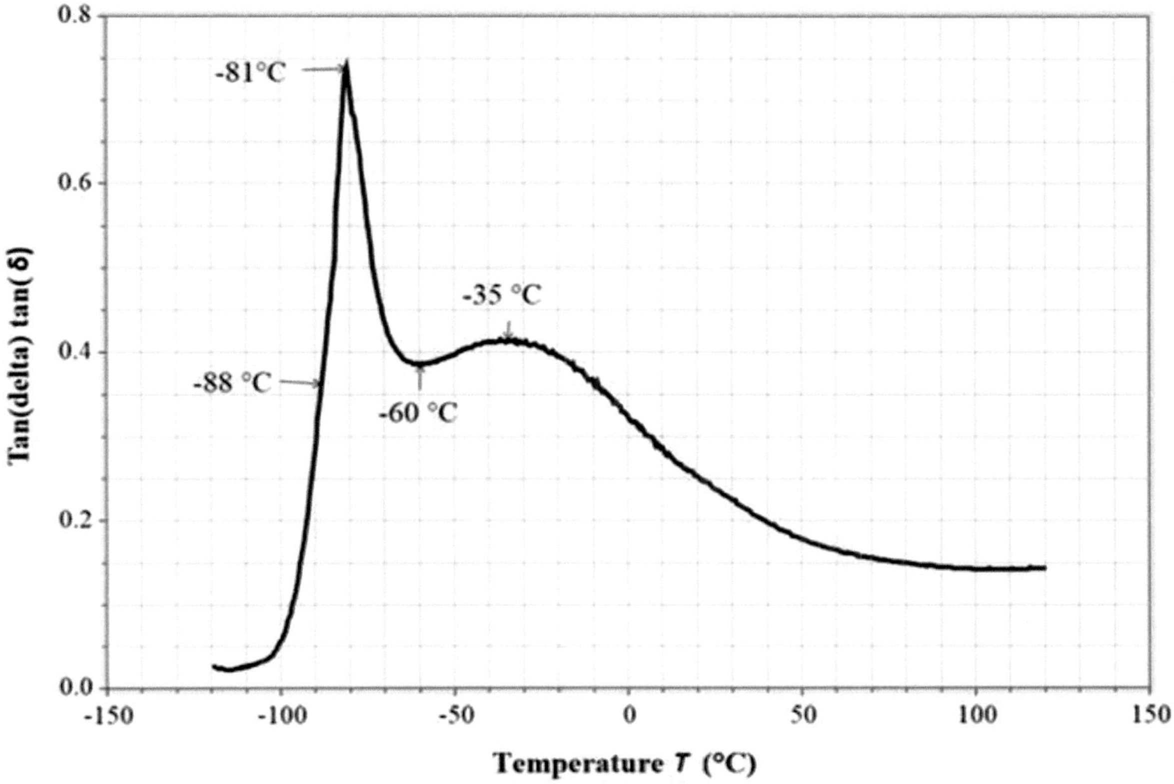

Fig. 2 Representative tan δ curve of filled HTPB composites [6]

Atomic force microscopy

AFM analysis was performed at room temperature (RT) and ambient pressures using a Jupiter XR AFM (Oxford Asylum Research) in the Fast Force Mapping (FFM) mode. Two cantilevers were used to investigate potential effects of a cantilever coating on the nanomechanical results. The AC160TS-R3 (uncoated lever) and the AC160TSA-R3 (lever coated with Cr/Au (5/65)) both consist of a 7 nm radius uncoated silicon tip, with a resonance frequency (f) of 300 (200–400) kHz and a spring constant (k) of 26 (8.4–7) N/m. Using this mode, the tip of the cantilever physically contacts the composite sample, allowing for nanomechanical information (such as adhesion and Young's modulus) to be derived in addition to standard topographical reconstruction of the surface. The total scan area for the present AFM images is 10 µm x 10 µm u, with individual line scans across a given interphase on a shorter scale (2-5 µm).

AFM sample preparation

For ideal measurements of the interphase using AFM, samples must be cross sectioned with extreme precision to expose a flat surface across the constituents [13–16]. Due to the highly compliant nature of HTPB binder, such uniform preparation has proven difficult to achieve with cryo-ultramicrotomy (see section 4.2 for discussion). Due to challenges cutting 73.49 wt. % loaded samples, all HTPB/glass bead composite samples for AFM analysis were 4.0 wt. % solids loading (1.5 volume %), nominally 25–40 µm size, with varied coating of the glass beads (uncoated or Sigmacote®). Each sample was razor cut to ~ 5mm (length) × 1mm (width) for ease of mounting in the cryo-microtome (Leica EM FC7). These samples were then sectioned for AFM analysis using two different microtome blades. A tungsten carbide knife (Delaware Diamond Knifes) was used to expose the desired cross section, followed by a diamond knife (DiATOME, cryo-AFM) that used a 50–100 nm feed to act as a "polishing" step for a smooth surface. Both composite samples and knives were held at −140°C (below the T_g of the HTPB system, determined to be -84 ° at 0.01 Hz in this work) with liquid nitrogen throughout the entire cutting process before being quickly warmed to RT under a nitrogen purge to mitigate the accumulation of unwanted condensation on the sample surface. Finally, the composite samples were secured to a magnetic AFM specimen disc with carbon tape for analysis.

Results and Discussion

DMA results

An overlay of the tan δ traces collected at 1 Hz is shown in figure 3a–d. Peak position for the first tan δ peak generally did not change across all systems investigated, indicating there is no appreciable change in T_g from the addition of filler. This observation is consistent with other studies on different filled composite systems that vary filler content and particle size [1, 9]. However, Bashir summarizes results of others that have reported the opposite and the shift in T_g appears to be rooted in the interaction strength between the particle and filler, which in turn can create dependencies on filler content and particle size [1].

The results also show that there is a height reduction in the first tan δ peak from the introduction of filler. The literature strongly suggests that the height of the tan δ peak is dependent on the particle size. The smaller the particles are, the lower the resultant tan δ peak will be. This can be attributed to more interfacial adhesion from the increase in surface area of filler [17–20]. Additionally, the tan δ peak height is believed to be dependent on the strength of the interfacial interaction. Stronger interactions result in the larger reductions of the tan δ traces [1, 18, 20].

The peak height of the first tan δ peak is plotted vs average particle size in figure 4. The coated glass beads always exhibited a higher peak value than their uncoated counterpart, which suggests that the Sigmacote® effectively reduced the particle-binder interaction strength. However, the trends in particle size did not agree with the expectations discussed above. For both the coated and uncoated systems, figure 4 shows that that the peak height diminished from the P4000 beads until the P0080 beads followed by an increase. The uncoated P0280 also traced lower than the P0170 where as the P0280-SC traced higher than the P0170-SC as expected.

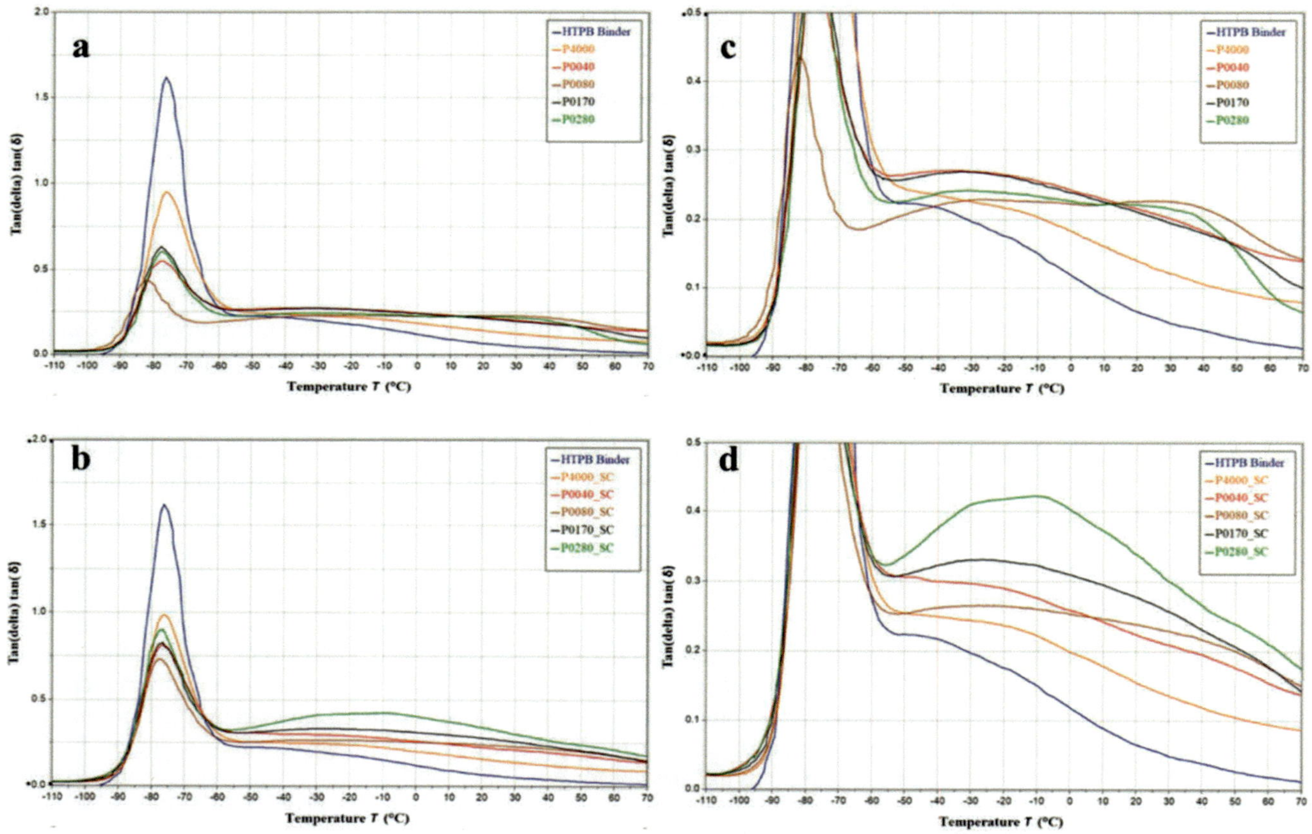

Fig. 3 1Hz tan δ vs temperature trace of HTPB binder and filled samples with **a)** uncoated glass beads and **b)** coated glass beads **c)** Rescaled tan δ vs temperature trace of uncoated glass beads **d)** Rescaled tan δ vs temperature trace of coated glass beads

One explanation for this unexpected behavior is the possible presence of agglomerates in the final samples. P4000, P4000-SC, P0040, and P0040-SC likely need longer and/or more aggressive mixing cycles to properly disperse the particles

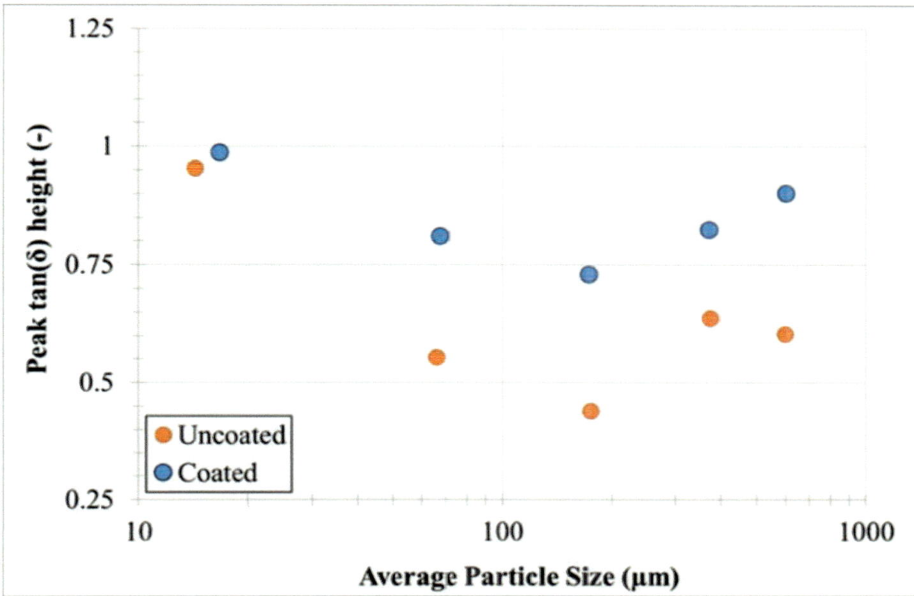

Fig. 4 Max tan δ height vs average particle size in HTPB/glass bead composites

in the binder. Without proper dispersion, a cluster of particles can effectively behave as a larger particle and reduce the overall surface area of the filler.

Additionally, not all samples exhibited a distinct secondary peak at warmer temperatures. Specifically, the P4000, P4000-SC, and P-0040-SC samples. With the suspected agglomeration issues, it is unknown if this is representative behavior or due to the lack of particle dispersion. The HTPB binder also appears to exhibit an underlying secondary tan δ peak. This suggests that there is a second relaxation process occurring in the binder alone, but it is impossible to determine if the introduction of filler affects the second relaxation process in the HTPB or creates a new relaxation process from creation of an interphase.

Furthermore, looking at the uncoated beads in figure 2b there is no clear trend in the trace heights above T_g. We would expect to see a correlation with particle size as discussed above, but any trends remain unclear even if the P4000 and P0040 samples are disregarded. However, a trend exists in the traces of the coated beads in figure 2d if sample P0040-SC is excluded. Sample P4000-SC is still suspected to be an outlier due to agglomeration concerns, but there appears to be a direct relationship between particle size and the height of the second tan δ peak as we should expect.

Comparing the coated and uncoated, the coated beads consistently exhibited a higher second tan δ peaks. The uncoated and coated P4000 and P0040 samples did not exhibit identifiable peaks, but the coated beads still showed higher tan δ traces beyond the T_g. Interestingly, it seems there is another tan δ peak for the P0280 and P0080 glass beads around 30 °C. The underlying cause is currently unknown but implies a potential third relaxation process for these systems.

EMG fitting

The tan δ curves were modelled after baseline correction with EMG functions shown in equation 1. The initial purpose of the EMG modelling was to quantify the peak heights, widths, and areas, and use T_{c_i} calculate the $E_a$of the relaxation processes. It was found that some systems could be modelled adequately with two EMG functions, while others required three. Optimization schemes for data sets requiring three EMG functions often yielded multiple analytical solutions for all parameters. Figure 5a shows an example of the range of EMG fitting solutions. Up to a 50 °C difference in T_{c_2} and T_{c_3}could occur when fits had multiple analytical solutions.

Additionally, some traces could be modelled adequately with two or three EMG functions as demonstrated in figure 5b. In these cases, the model with three EMG functions did not appear to accurately assign T_{c_2} to the second tan δ peak due to over fitting, but the models with two EMG function did. Lastly, most of the second tan δ peak height differences were erased with the baseline correction function that minimized differences for A_i between samples. Analysis of the EMG function parameters and T_{c_i} values for E_a calculations is not reported for reasons just mentioned.

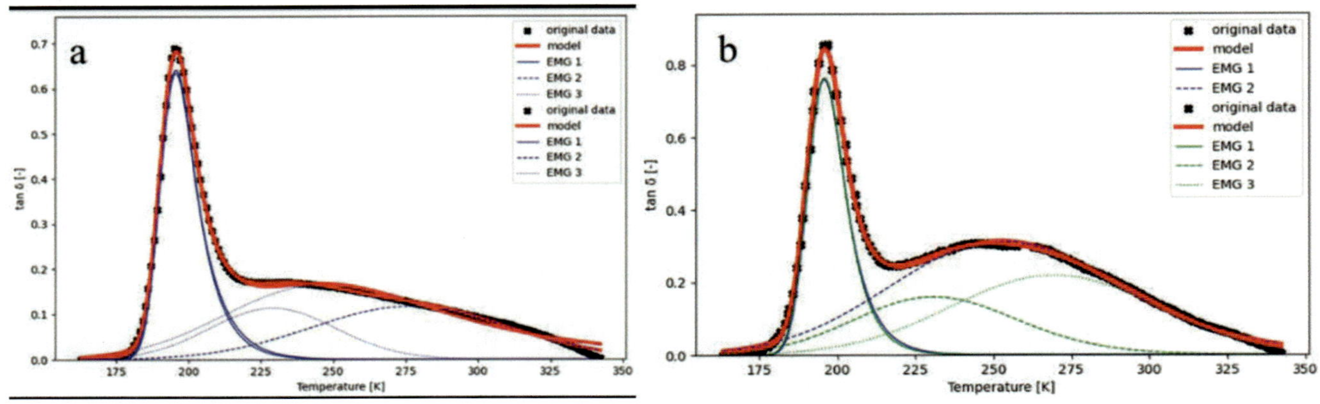

Fig. 5 **a)** Example of EMG model fit resulting in multiple analytical solutions for P0080-SC **b)** P0280-SC with fitted with two and three EMG functions

Activation energy calculations

Due to the issues with the EMG fitting, E_a calculations were completed with the peak heights directly from the tan δ traces rather than the T_{c_i}. Peak temperatures are reported in table 3 below. Values are not reported when a material or test condition did not result in any identifiable peaks. 10 Hz data is also not reported for the second tan δ peak due to phase angle issues with data above Tg. Strain % increases within the LVR can be employed in the future to mitigate this issue when testing above the binder T_g [4].

The results suggest that introduction of the filler increases the E_a of the bulk polymer T_g and the E_a of the second tan δ peak when comparing neat binder to all the filled samples. Looking specifically at the E_a for the T_g peak, no obvious trends exist between E_a and the filler particle size, or the E_a of coated vs uncoated beads. Data fitting for the second tan δ peak is relatively poor due to the lack of frequencies available. Including additional frequencies in future testing will allow more accurate E_a calculations. Despite the lack of data, there appears to be a difference between the E_a of coated and uncoated samples. Again, excluding P4000, P4000-SC, P0040, and P0040-SC samples, the remaining materials consistently exhibited a lower E_a in the coated glass beads vs the uncoated counterparts. The reduction in E_a indicates a weaker interaction with the binder [12].

Particle-binder interaction parameter

The particle-matrix interaction parameter, *A*, was first proposed by Chua and adapted by Kubat *et al.* [17, 20]. The equation provided by Kubat *et. al.* is shown in equation 3 and is based on the energy loss at the interface. In equation 3, v_f is the volume fraction of filler, and $tan\,(\delta_c)$ and $tan\,(\delta_m)$ are the composite tan δ and binder tan δ traces, respectively, as a function of temperature. The parameter *A* discussed here is different than the model parameter A_i discussed previously in equation 1. The parameter *A* has an inverse relationship with the polymer-filler interaction, meaning lower *A* values indicate a stronger particle-binder interaction. Calculations of the interaction parameter may shed light on interphase characteristics when coupled with AFM results.

$$A = \frac{1}{1 - v_f} \bullet \frac{tan\,(\delta_c)}{tan\,(\delta_m)} - 1 \tag{3}$$

The interaction parameter *A* is plotted vs temperature in figure 6 for the 1 Hz data sets. All samples follow a similar shape and can be generally broken down into three regimes. In the first regime, *A* increases slightly from -110 °C to -95 °C which indicates a weakening interaction strength as the temperature approaches T_g. Second, *A* decreases from -95 °C to the -75 °C (in the T_g range at 1 Hz) and may be attributed to increased molecular mobility in the binder which increases viscosity and friction at the interface [21, 22]. Third, *A* generally increases from -75 °C until the end of the test where the binder particle interaction decreases as molecular mobility continues to increase [21, 22]. Some samples show a decrease in *A* at warmer temperatures, with the P0280 samples showing a significant increase in the particle-binder interaction. The underlying mechanism for this decrease of *A* at warmer temperatures is currently unknown.

The difference in the *A* parameter between coated and uncoated glass beads is also apparent across all samples above their T_g. First, every coated sample shows a higher *A* parameter, again suggesting that the interaction strength is weaker than the uncoated counterpart. Next, looking at uncoated samples, the *A* parameter does not exhibit many differences across

Table 3 Activation Energy calculations for T_g and second tan δ peak

T_g (°C) values at each test frequency						
Sample	0.01 Hz	0.1 Hz	1 Hz	10 Hz	E_a (kJ/mol)	R^2
HTPB	-83.76	-80.25	-76.26	-70.75	169.16	0.993
P4000	-82.92	-79.9	-76.16	-71.15	186.46	0.991
P4000-SC	-82.98	-79.99	-75.98	-71.00	182.44	0.991
P0040	-85.16	-80.82	-76.99	-72.75	175.47	0.999
P0040-SC	-83.97	-81.03	-76.97	-71.96	179.84	0.991
P0080	-85.39	-81.95	-77.97	-73.58	181.70	0.999
P0080-SC	-84.07	-80.97	-77.94	-73.04	198.12	0.989
P00170	-85.07	-80.57	-77.69	-73.16	185.11	0.993
P0170-SC	-83.49	-81.47	-76.55	-71.89	179.22	0.976
P0280	-84.24	-81.03	-77.3	-72.09	179.19	0.991
P0280-SC	-84.09	-80.80	-76.79	-73.21	197.23	0.999
2nd tan δ peak temperature (°C) at each test frequency						
Sample	0.01 Hz	0.1 Hz	1 Hz	10 Hz	Ea (kJ/mol)	R^2
HTPB	-58.77	-52.91	-45.13	-	136.66	0.996
P4000	-	-	-		-	-
P4000-SC	-50.9	-44.00	-		141.30	1.00
P0040	-40.52	-34.97	-35.82		279.28	0.621
P0040-SC	-52.03	-49.98	-48.05		478.58	0.999
P0080	-31.76	-26.85	-26.82		347.70	0.755
P0080-SC	-41.02	-35.03	-26.05		145.41	0.991
P00170	-38.7	-33.82	-31.97		300.37	0.934
P0170-SC	-35.28	-32.32	-26.95		262.63	0.976
P0280	-38.36	-35	-31.97		338.81	0.999
P0280-SC	-30.05	-27.79	-10.39		104.59	0.841

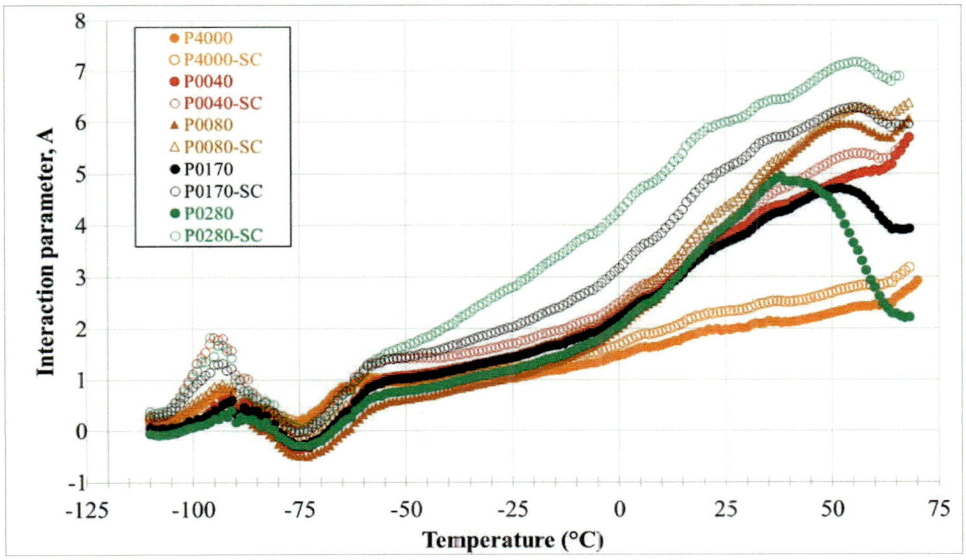

Fig. 6 Interaction parameter A vs temperature where parameter A has an inverse relationship with particle-binder interfacial strength

particle sizes until approximately -10 °C, where the P4000 sample continues increasing at the same slope, and all other samples exhibit a slope increase. There are not many differences between the P0040, P0080, P0170, and P0280 samples until they exhibit their respective decreases at warmer temperatures.

Finally, examining the coated samples, all samples have a similar *A* parameter below T_g, but distinct differences emerge above T_g. Smaller coated particles generally showed lower *A* values than the larger coated particles. Samples also displayed an inverse relationship between the slope of the *A* parameter above T_g. This implies that smaller particles experience higher particle-binder interactions and are not as sensitive to changes in the interaction at temperatures above T_g. While the larger particles see a faster increase in the *A* parameter across temperature increases.

Other approaches are available to calculate various interaction parameters or specific activation energies of the interphase. However, all these approaches require systematic changes in the volume % to calculate these values [11, 23]. This type of data is not available yet for this work but may serve as an alternative in the future to characterize interphase properties via DMA.

AFM Results

Highly loaded and/or large particle sized glass bead samples, as used in the DMA experiments, were found to pose a great risk of damaging the microtome knives, thus damaging the binder and causing large variation in height between constituents. During initial attempts to microtome samples with P0280 glass beads (602 μm average particle size), the tungsten carbide knife chipped and was unable to section the bead. In an attempt to preserve the knives operational life, only smaller glass bead sizes were used moving forward. Figure 7 shows an optical image of the surface of the 50 volume % P4000 sample prepared via cryo-microtome. Although the glass beads were cut in half, the binder remains rough and damaged—visualized as unfocused optically, opaque, no discernable features that result from cutting (seen as elongated strides on the glass beads)—which is likely result of a damaged/dull knife unable to section the sample surface properly. To increase the cut quality and protect the knives from further damage, 4 wt. % particle loadings were used as they minimize the frequency of particle contacts with the knife.

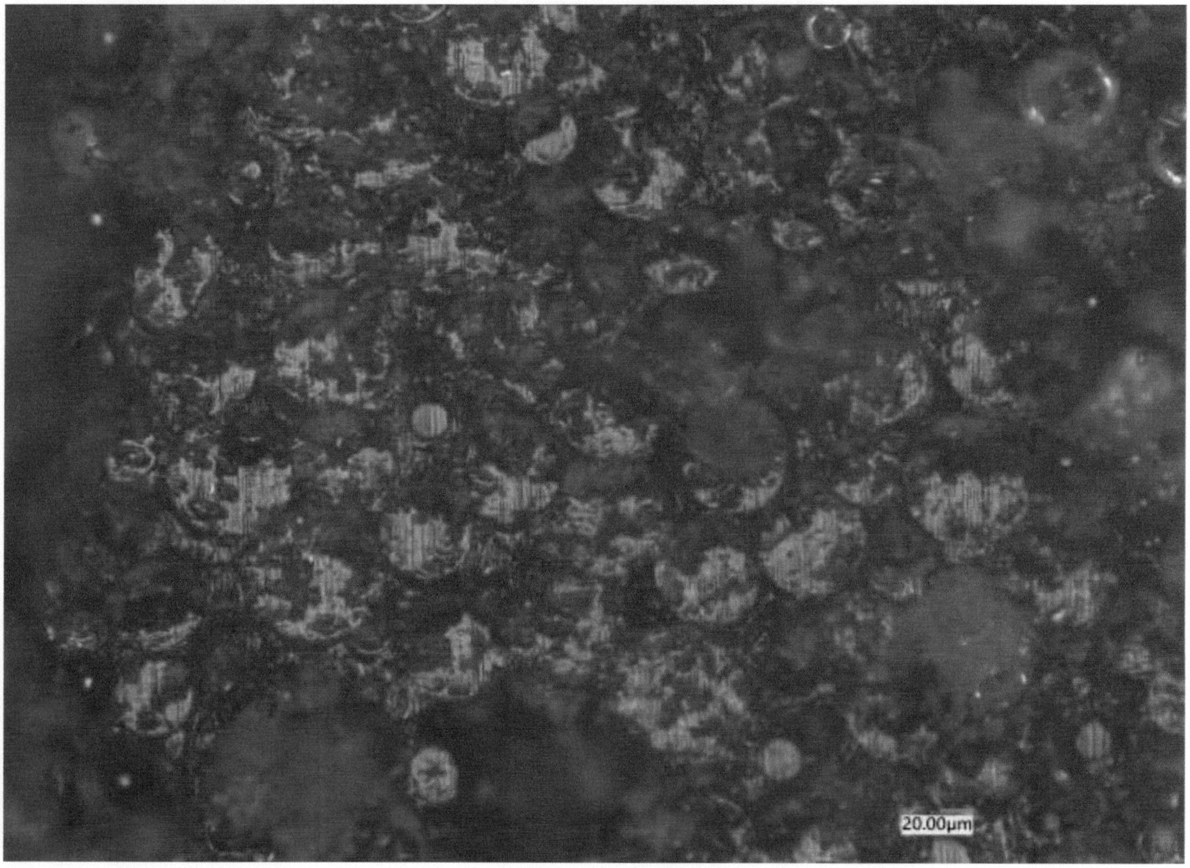

Fig. 7 Optical image of the surface of sample P4000 sample at 50 volume % solids loading prepared via cryo-microtome

When the knives did not sustain appreciable damage during the cutting process, cutting thin slices from the surface was used to mitigate large variations in height across the surface and between constituents. However, when the binder is as compliant as HTPB and must be cooled to section, another consequence arose. We observe that the embedded particle—having the appearance a pristine flat cut optically at cryogenic temperatures—appeared to be protruding from the binder once the sample was brought to RT (see the top (red) trace of figure 8 for reference). We believe this phenomenon is a direct result of HTPB's relatively large coefficient of thermal expansion to that of soda-lime silica glass of HTPB as it warms to RT, thus altering the prepared surface [24]. Other surface preparation approaches—cryo- and RT hand polishing and cryo- and RT broad beam ion milling—have been attempted to overcome this unforeseen consequence. For each mitigation strategy, we observe issues. Particle pull out (from hand polishing), charring of the HTPB (from ion milling), and metal re-deposition (from ion milling) significantly alter the surface and, in turn, irreversibly alter the interphase. At the time of this study, we are continuing to investigate other cross sectioning techniques (such as oscillating diamond knives, diamond wire saw, etc.) that will not induce damage and minimize the observed thermal expansion issue, while still preparing a flat surface for high quality AFM measurements.

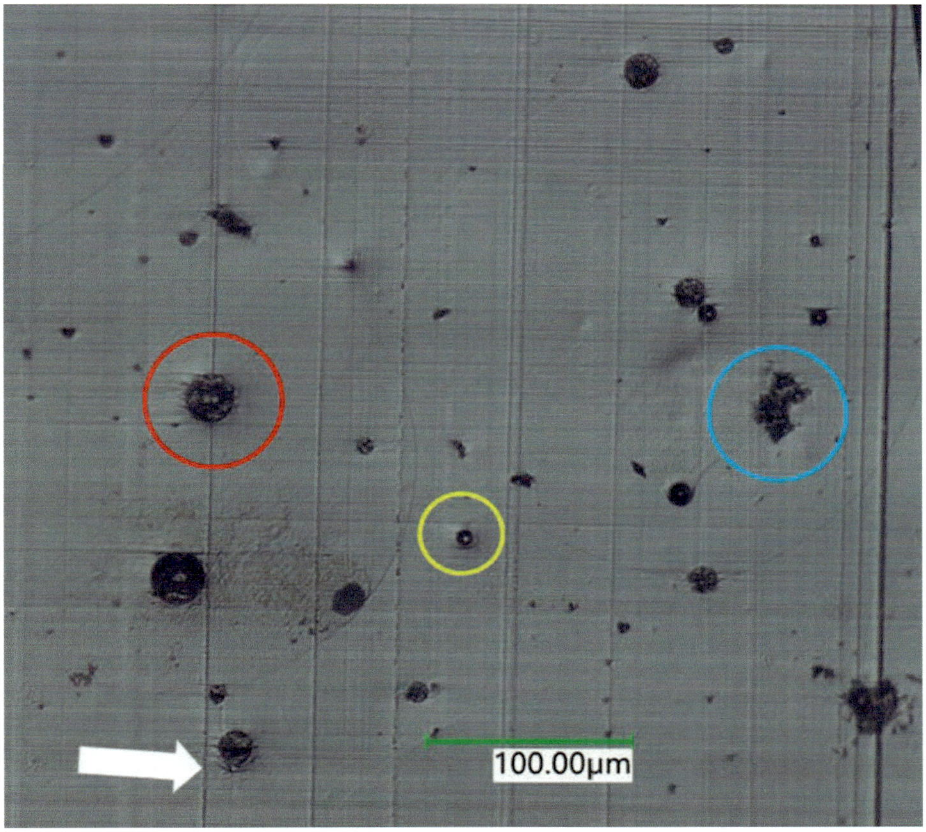

Fig. 8 Optical image of the surface of P4000 (4 wt. %) composite sample prepared via cryo-microtome. Cutting direction is from top to bottom of the image. Elongated vertical lines running through the sample are knife marks, an artifact based on knife damage. Colored circles and arrow highlight interesting features of the cross section, see main text

Figure 8 shows a representative optical image of the cryo-microtome P4000 (4 wt. %) surface used for AFM analysis. We note key features of the cross sectioned surface to highlight the difficulties of sample preparation, and the obstacles faced in finding a viable bead/interphase to analyze. The cutting direction is from top to bottom of the image. The red circle in figure 8 shows a glass bead that is cut in half (and appears slightly raised from the HTPB binder) but does not exhibit the same knife marks as the surrounding HTPB sample (elongated vertical lines). Such an artifact can be indicative of a fractured glass bead instead of one that is cleanly cut. Highlighted in yellow in figure 8, we observe a vacancy in the HTPB where a glass bead was pulled out of the sample. The blue circle in figure 8 highlights a conglomerate of smaller glass beads that presents as a non-spherical cluster. Finally, we can observe a 'lip' behind some of the embedded beads (shown with the white arrow in figure 8) that correlates with the 'back' of the bead in terms of cutting direction. All of these artifacts allude to damage imparted by the cutting process, which will likely affect the measured interphase width.

Nevertheless, we attempt AFM analysis on coated and uncoated P4000 composites at 4 wt. % loading. Two glass beads were identified in the P4000-SC sample and are denoted as bead A and bead B. Two glass beads are also identified in the

P4000 sample and denoted as beads C and D. The nanomechanical properties of a sample measured using AFM consist of the height, adhesion, and Young's modulus. These properties can be presented in two configurations, as shown in figure 9 as 3D/2D color map images and figure 10 as a plot. For the AFM analysis herein, we define adhesion as the force needed to disengage the tip of the cantilever from the surface of the sample and the Young's modulus as the measure of the sample's resistance to elastic deformation under a given load, calculated using the known tip geometry of the selected cantilever and assuming Hertzian contact mechanics. The adhesion and Young's modulus values reported are relative rather than absolute because 1) it is historically difficult to validate such analytical approach across constitutes of drastically different properties and height and 2) the main goal of AFM analysis is to measure the width of the interphase (a measure of how far, moving away from the glass bead, until the binder properties are representative of the bulk).

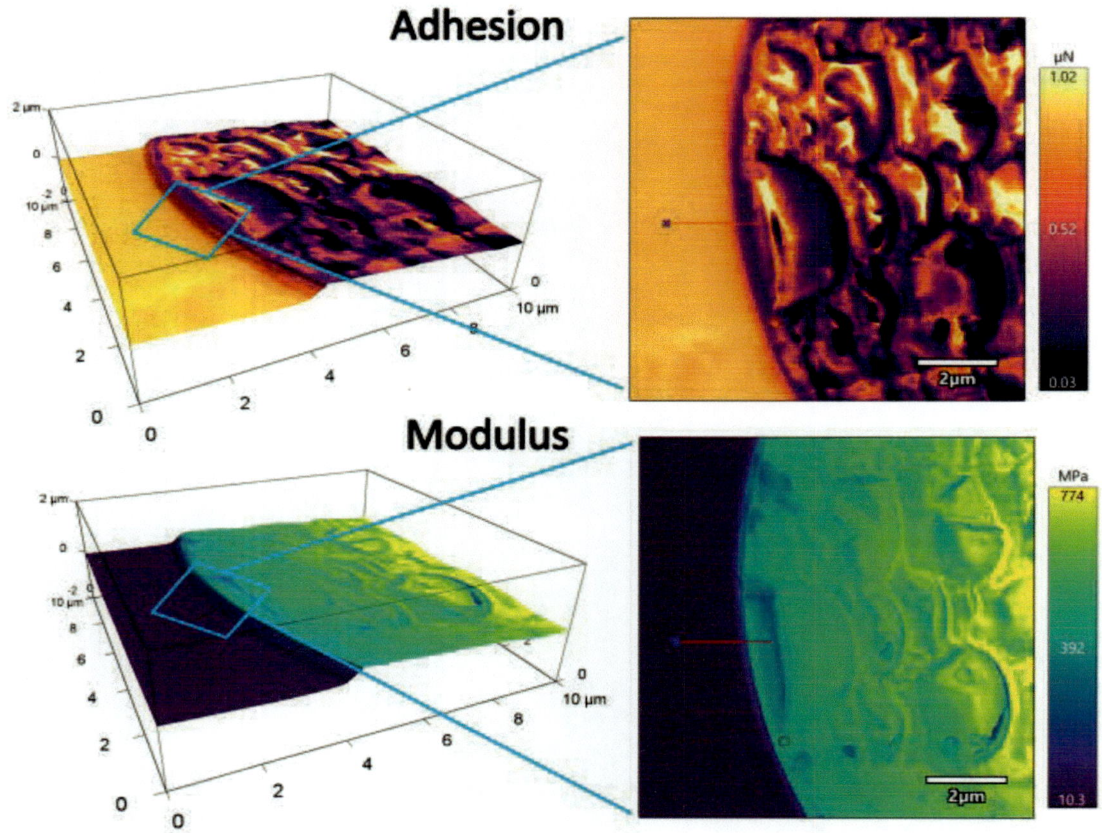

Fig. 9 3D images of P4000-SC (~30 μm, bead A) sample topography (left) obtained over a 10 μm × 10 μm FFM scan overlaid with a color map representing the adhesion (top, Inferno color map) and Young's modulus (bottom, Viridis color map). 2D zoomed-in images (right) of adhesion (top) and modulus (bottom) interphase, where the solid red line represents the line scan across the interphase. The HTPB binder is on the left side of each image and the glass bead is on the right side

Highlighted in Collinson et al. 2021 (and references therein), changes in sample height can induce non-ideal tip-surface interactions that can in turn lead to 'dead-zones' where the information gathered becomes non-quantitative [14]. It is extremely difficult to detangle whether the sudden change in height causes the change in nanomechanical properties, or if the sudden change in nanomechanical properties causes artifacts in the topography resulting from different indentation depths. Acknowledging the likelihood of dead-zones that could result from non-flat surface preparation, we report tentative interphase widths as we continue to investigate best practices and mitigate undesirable artifacts of multi-component systems with starkly different properties.

The 3D images (figure 9, left) visualize the topography of a given binder/bead region (bead A) of a P4000-SC sample across a 10 μm × 10 μm scan area, overlaid with the adhesion (figure 9, top panel) or Young's modulus (figure 9, bottom) information. The 2D images (figure 9, right) represent a smaller portion of the sample surface, more easily depicting the color gradient moving from the bulk HTPB (represented as orange in the adhesion and dark blue in the modulus) to the bulk uncoated glass bead (represented as purple in the adhesion and green/yellow in the modulus) for both adhesion and Young's modulus. The uneven topography of the glass bead and the height difference between constituents is easily visualized using this representation of data.

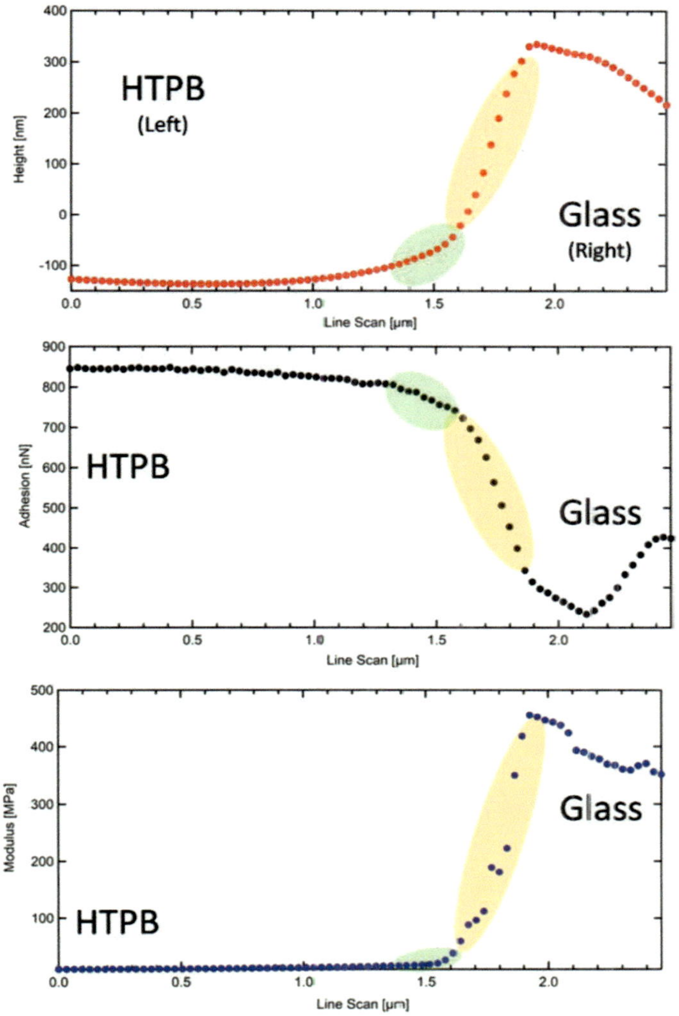

Fig. 10 Line scan across the interphase region of a different P4000-SC (4 wt. %) glass bead (bead A) showing the evolution in height (top, red circles), adhesion (middle, black circles), and modulus (bottom, blue circles) of the sample. The green highlighted areas show a "loosely bound interphase" region and the yellow highlighted areas show a "tightly bound interphase" region (see main text for explanation)

To represent the color gradient data from figure 9 graphically, a line scan (red line in 2D images of figure 10) was performed, allowing for the evolution of the height, adhesion, and Young's modulus to be plotted against the line scan distance, shown in figure 10.

Two distinct transition regions are recognized when analyzing the adhesion and modulus line scan for bead A (middle and bottom traces of figure 10, respectively). The bulk HTPB is a soft and compliant binder, which translates to a large adhesion value (~850 nN) and a small modulus value (10s of MPa). As the tip moves from the bulk binder towards the coated glass bead A, the adhesion and modulus values begin to deviate from those observed in the bulk, which can be referred to as the 'loosely bound' interphase region (highlighted in green in figure 10). In this region—that spans from ~530 nm to ~190 nm away from coated glass bead A—the adhesion decreases ~150 nN and the modulus increases ~70 MPa. The change in height over this "loosely bound interphase" region is ~110 nm. As the tip continues to move towards the coated glass bead A, there is another observable change in adhesion and modulus, highlighted in yellow in figure 10, that can be considered the "tightly bound interphase" region. From ~190 nm away from coated glass bead A there is a steep drop in the adhesion value (an additional ~320 nN) and an even more drastic increase in the Young's modulus (increasing ~260 MPa) until the values converge on the average bulk value of coated glass bead A. The height change for this "tightlt bound interphase" region is ~260 nm. From these results, the total interphase width of the bead A in the P4000-SC (4 wt. %) sample is ~ 530 nm.

Bead B from P4000-SC (4 wt. %) in the same cross section was investigated. A line scan was performed, showing evolution of the height, adhesion, and Young's modulus of Sigmacote glass bead B in HTPB in figure 11. Again, there are two distinct transition regions observed as the scan moves from bulk HTPB towards the embedded bead. Both the adhesion and modulus values of bulk HTPB remain consistent between scans (~ 850 nN and 10s of MPa, respectively). However, we begin to see deviations from bulk behavior beginning ~ 2.4 μm away from the glass bead. From ~ 2.4 μm to 600 nm away from the bead, the adhesion drops ~300 nN and the modulus increases ~10 MPa over a height change of ~140 nm in the "weakly bound interphase" region, highlighted in green in figure 11. From 600 nm away to the edge of the bead, we again observe a more drastic change in nanomechanical properties in the "tightly bound interphase" region highlighted in yellow. We observe a decrease in adhesion of ~ 300 nN and an increase of ~150 MPa in modulus across a ~190 nm height change. From the results in figure 11, the total interphase width of the investigated bead in the coated glass bead and HTPB sample is ~ 2.4 μm, a stark difference from the previously investigated bead in figure 10. The range of interphase widths measured using AFM for the beads in P4000-SC (4 wt. %) are summarized in Table 4.

Next, we investigate uncoated glass beads in the P4000 (4 wt. %) sample from the same cross section using the same methodology. figure 12 shows the evolution of the height, adhesion, and Young's modulus across the interphase region of the

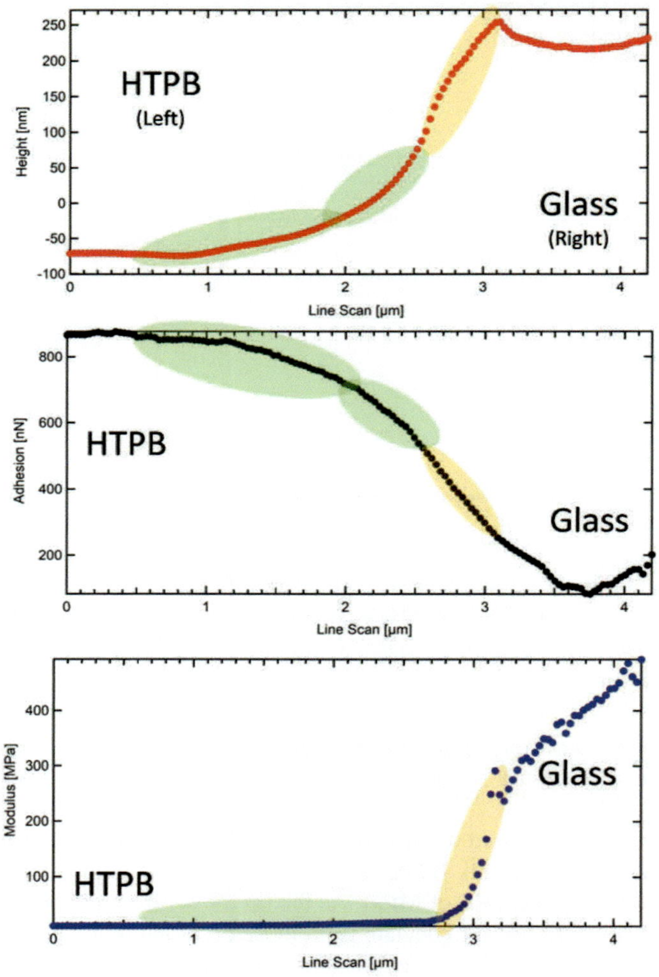

Fig. 11 Line scan across the interphase region of a different P4000-SC (4 wt. %) glass bead (bead B) showing the evolution in height (top, red circles), adhesion (middle, black circles), and modulus (bottom, blue circles) of the sample. The green highlighted areas show a "weakly bound interphase" region and the yellow highlighted areas show a "tightly bound interphase" region (see main text for explanation)

glass bead sample (denoted bead C). When analyzing this interphase region, we observe a deviation from bulk adhesion and modulus values beginning ~440 nm away from the uncoated glass bead C. The adhesion value decreases ~130 nN moving from ~440 nm to ~95 nm away from the interface, while the modulus increases ~45 MPa. We again define this change from bulk properties as the 'weakly bound interphase' region, highlighted in green in figure 12. As the tip moves closer to the embedded bead, a second discernable change in the adhesion and modulus values is realized. From ~95 nm away from uncoated bead C to the bead edge in the "tightly bound interphase" region, the adhesion value decreases ~200 nN while the modulus increases ~290 MPa. The total change in height over both the "weakly bound interphase" and 'tightly bound interphase" regions was nearly 450 nm, and the total interphase width is measured to be ~ 450 nm for uncoated bead C in this sample.

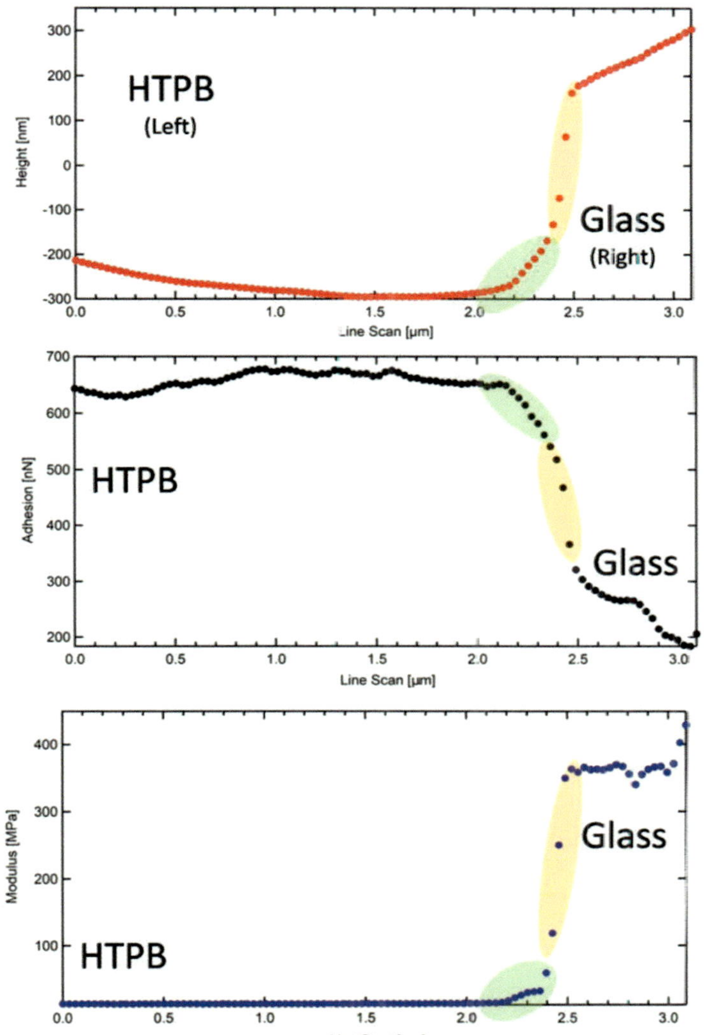

Fig. 12 Line scan across the interphase region of a P4000 (4 wt. %) glass bead (bead C) showing the evolution in height (top, red circles), adhesion (middle, black circles), and modulus (bottom, blue circles) of the sample. The green highlighted areas show a "weakly bound interphase" region and the yellow highlighted areas show a "tightly bound interphase" region (see main text for explanation)

Finally, we investigate another bead within the same cross section of the P4000 (4 wt. %) sample, defined as bead D. Deviations from the bulk adhesion and modulus values begin ~ 2.3 µm away from uncoated glass bead D, shown in green in figure 13. From ~2.3 µm to 400 nm away from the uncoated glass bead D the adhesion value decreases ~250 nN while the modulus increases ~11 MPa. Moving from 400 nm away to the edge of the uncoated bead D, there is another stark change in the adhesion and modulus values over a 350 nm height gradient. The adhesion value drops ~230 nN and the modulus

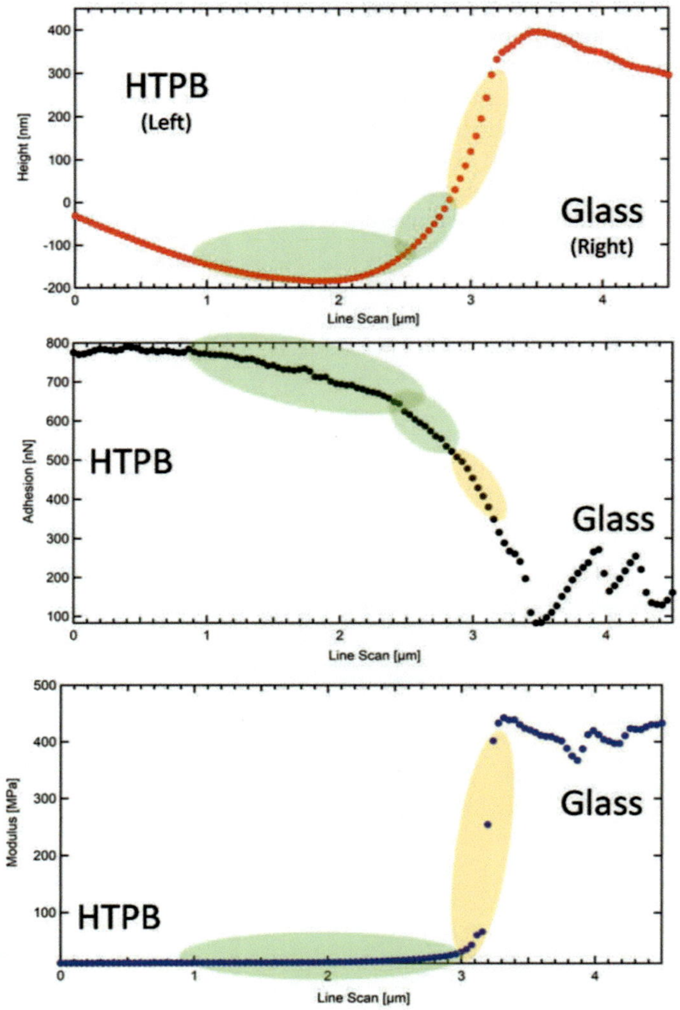

Fig. 13 Line scan across the interphase region of a different P4000 (4 wt. %) glass bead (bead D) showing the evolution in height (top, red circles), adhesion (middle, black circles), and modulus (bottom, blue circles) of the sample. The green highlighted areas show a "weakly bound interphase" region, and the yellow highlighted areas show a "tightly bound interphase" region (see main text for explanation)

increases ~380 MPa before both values plateau to the bulk values of the uncoated glass bead measured under the present conditions. The total interphase width measured for uncoated glass bead D in HTPB (over a large deviation in height over 500 nm) is ~2.3 μm, summarized in Table 4.

The results of the interphase width range for the 25–40 μm coated and uncoated glass beads in their respective samples are presented in Table 4. From this AFM study on embedded glass beads in HTPB, there is no observed consistency of the interphase width between beads of the same coating within a given cross section. Additionally, there is no observed difference in the interphase width between a coated glass bead and an uncoated glass bead in HTPB. We hypothesize the AFM results presented are dominated by a large height difference between constituents as opposed to a true representation of the interphase width. Nevertheless, it is important to highlight how AFM analysis of the interphase region is heavily influenced by the sample preparation method which, in turn, is heavily reliant on the binder and embedded particle chosen.

Additional bead sizes were not characterized at the time of this publication due to time constrains. Given the challenges cutting larger bead sizes, it is unlikely that the other samples would have yielded improved results. The variability of the interphase width measurements for the coated and uncoated P4000 (4 wt. %) samples, and a lack of data for other bead sizes prevents any correlation of the DMA results to interphase properties at this time. These challenges highlight a need to improve AFM sample preparation techniques for follow on investigations.

Table 4 Summary of interphase widths for P4000 (4 wt. %) and P4000-SC (4 wt. %) from AFM analysis

Glass Bead Coating	Average Particle Size (μm)	Interphase Width Range (μm)
Coated	16.75	0.5–2.4
Uncoated	14.35	0.4–2.3

Conclusions

This study investigated the particle-binder interphase regions in HTPB/glass bead composites with DMA and AFM and sought to correlate data between the two techniques. Glass beads were employed as the filler in a HTPB binder system and were varied from an average particle size of approximately 16 μm to 600 μm at a fixed volume % of filler. Glass beads were also coated with Sigmacote® to act as a shielding agent to weaken the particle-binder interaction strength.

DMA tan δ traces showed two clear peaks for most of the samples evaluated which indicated two separate relaxation process occurring within the material. The first peak at the colder temperature is from the T_g of the HTPB. The particle size and the coating on particles did not show any appreciable change to the T_g, but did result in differences in peak height. Investigation of the second tan δ peak did not explicitly prove any correlates to relaxation mechanism solely within the interphase, but these results support the argument that the interphase characteristics could affect the peak's size and shape. General trends in the tan δ vs temperature plots show that the particle size and Sigmacote® affect the peak heights. Outliers in the reported trends are suspected to be due to poor mixing during sample generation. Activation energy and interaction parameter calculations also suggest that changing the particle size and particle surface chemistry will affect the particle-binder interaction strength and subsequently, the interphase.

Correlating DMA data to AFM measurements was not possible at this time due to the challenges associated with AFM sample preparation. Due to the compliant nature of HTPB binder, the hardness of the glass beads, and the differences in the binder's and glass beads' respective coefficients of thermal expansion, it was difficult to achieve a pristine cross-section cut. AFM measurements do suggest the presence of "weakly bound interphase" and "tightly bound interphase" regions, however, there is a significant variability in the measurement. This highlights the need to improve sample preparation techniques to permit any correlation of AFM interphase measurements to DMA data in future work.

References

1. Bashir, M. A. (2021). Use of dynamic mechanical analysis (DMA) for characterizing interfacial interactions in filled polymers. *Solids*, *2*(1), 108–120. https://doi.org/10.3390/solids2010006
2. Adicoff, A., & Lepie, A. H. (1970). Effect of tensile strain on the use of the WLF equation. *Journal of Applied Polymer Science*, *14*(4), 953–966. https://doi.org/10.1002/app.1970.070140406
3. Azoug, A., Nevière, R., & Constantinescu, A. (2015). Molecular origin of the influence of the temperature on the loss factor of a solid propellant. *Propellants, Explosives, Pyrotechnics*, *40*(4), 469–478. https://doi.org/10.1002/prep.201400060
4. Bohn, M. A., Lemos, M. F., Mussbach, G. (2017) *Influence of concentration, type and particle size of fillers on the dynamic mechanical behaviour of elastomeric HTPB binder*. Fraunhofer ICT (International Annual Conference) 2017, Karlsruhe, Germany.
5. Gańczyk-Specjalska, K., & Magnuszewska, P. (2020). An analysis of the mechanical properties of HTPB-propellants using DMA. *Materiały Wysokoenergetyczne / High Energy Materials*, 81–91. https://doi.org/10.22211/matwys/0192
6. Tramell, J., Bohn, M., & Gerber, P. (2024, June 25-28) *Reproducibility and Applicability of Time-Temperature Superposition for Polymer Bonded Explosives Tested with Torsional Dynamic Mechanical Analysis*. 53rd International Annual Conference of the Fraunhofer ICT on 'Energetic Materials – Structure and Properties', Karlsruhe, Germany.
7. Cerri, S., Bohn, M. A., Menke, K., & Galfetti, L. (2013). Aging of HTPB/AL/AP rocket propellant formulations investigated by DMA measurements. *Propellants, Explosives, Pyrotechnics*, *38*(2), 190–198. https://doi.org/10.1002/prep.201200186
8. Seyidoglu, T., & Bohn, M. A. (2020). Effects of four isocyanates and four plasticizers on the thermomechanical and tensile properties of hydroxyl-terminated polybutadiene elastomers and the effect of solid particle filling. *Journal of Applied Polymer Science*, *138*(18). https://doi.org/10.1002/app.50362
9. Tsagaropoulos, G., & Eisenberg, A. (1995). Dynamic mechanical study of the factors affecting the two glass transition behavior of filled polymers. similarities and differences with random ionomers. *Macromolecules*, *28*(18), 6067–6077. https://doi.org/10.1021/ma00122a011
10. Bohn, M. A., Musbach, G., & Cerri S. (2012, June 26-29) *Influences on the Loss Factor of Elastomer Binders and its Modeling*. 43rd International Annual Conference of the Fraunhofer ICT on 'Energetic Materials – Synthesis, Characterisation, Processing', Karlsruhe, Germany. ISSN 0722-4087.
11. Li, R., Xu, Y., Zhang, K., Wu, Z., Ma, N., & Ouyang, X. (2024). A novel computational model based on molecular chain dynamics: Quantification and evaluation of interfacial interactions in polymer-based composite materials. *Composites and Advanced Materials*, *33*. https://doi.org/10.1177/26349833241309350

12. Lemos, M. F., Mussbach, G., & Bohn, M. A. (2017). Evaluation of filler effects on the dynamic mechanical behavior of HTPB-elastomer used as binder in exemplary composite formulations. *Journal of Aerospace Technology and Management*, *9*(3), 379–388. https://doi.org/10.5028/jatm.v9i3.795
13. Cheng, X.; Putz, K. W.; Wood, C. D.; Brinson, L. C. Characterization of Local Elastic Modulus in Confined Polymer Films via AFM Indentation. *Macromol. Rapid Commun.* **2015**, *36* (4), 391–397. https://doi.org/10.1002/marc.201400487.
14. Collinson, D. W.; Sheridan, R. J.; Palmeri, M. J.; Brinson, L. C. Best Practices and Recommendations for Accurate Nanomechanical Characterization of Heterogeneous Polymer Systems with Atomic Force Microscopy. *Progress in Polymer Science* **2021**, *119*, 101420. https://doi.org/10.1016/j.progpolymsci.2021.101420
15. Tian, C.; Chu, G.; Feng, Y.; Lu, Y.; Miao, C.; Ning, N.; Zhang, L.; Tian, M. Quantitatively Identify and Understand the Interphase of SiO2/Rubber Nanocomposites by Using Nanomechanical Mapping Technique of AFM. *Composites Science and Technology* **2019**, *170*, 1–6. https://doi.org/10.1016/j.compscitech.2018.11.020
16. Monclus, M. A.; Young, T. J.; Di Maio, D. AFM Indentation Method Used for Elastic Modulus Characterization of Interfaces and Thin Layers. *J Mater Sci* **2010**, *45* (12), 3190–3197. https://doi.org/10.1007/s10853-010-4326-6.Dynamic Mechanical Analysis Studies of the Interphase - PING SENG CHUA
17. Bashir, M. A., Jakobsen, M. G., & Farstad, V. B. (2020). The effect of extender particle size on the glass transition temperature of model epoxy coatings. *Polymers*, *12*(1), 196. https://doi.org/10.3390/polym12010196
18. Hayeemasae, N., Soontaranon, S., & Masa, A. (2024). Comparative investigation of nano-sized silica and micrometer-sized calcium carbonate on structure and properties of natural rubber composites. *Polymers*, *16*(8), 1051. https://doi.org/10.3390/polym16081051
19. Robertson, C. G., Lin, C. J., Rackaitis, M., & Roland, C. M. (2008). Influence of particle size and polymer−filler coupling on viscoelastic glass transition of particle-reinforced polymers. *Macromolecules*, *41*(7), 2727–2731. https://doi.org/10.1021/ma7022364
20. Kubát, J., Rigdahl, M., & Welander, M. (1990). Characterization of interfacial interactions in high density polyethylene filled with glass spheres using dynamic-mechanical analysis. *Journal of Applied Polymer Science*, *39*(7), 1527–1539. https://doi.org/10.1002/app.1990.070390711
21. Lin, Cong-mei, Liu, S., Wen, Y., Liu, J., He, G., Zhao, X., Yang, Z., Ding, L., Pan, L., Li, J., & Guo, S. (2022). Sandwich-like interfacial structured polydopamine (pda)/wax/PDA: A novel design for simultaneously improving the safety and mechanical properties of highly explosive-filled polymer composites. *Energetic Materials Frontiers*, *3*(4), 189–198. https://doi.org/10.1016/j.enmf.2022.03.003
22. Lin, Congmei, Yang, X., He, G., Wen, Y., Qian, W., Liu, R., Liu, S., Gong, F., Zhang, J., Zeng, C., Yang, Z., Chen, R., & Guo, S. (2023). Mussel-inspired interfacial reinforcement of thermoplastic polyurethane based energetic composites. *Composites Science and Technology*, *232*, 109875. https://doi.org/10.1016/j.compscitech.2022.109875
23. Ashida, M., Noguchi, T., & Mashimo, S. (1985). Effect of matrix's type on the dynamic properties for short fiber-elastomer composite. *Journal of Applied Polymer Science*, *30*(3), 1011–1021. https://doi.org/10.1002/app.1985.070300311
24. Dossi, E.; Earnshaw, J.; Ellison, L.; Rabello Dos Santos, G.; Cavaye, H.; Cleaver, D. J. Understanding and Controlling the Glass Transition of HTPB Oligomers. *Polym. Chem.* **2021**, *12* (17), 2606–2617. https://doi.org/10.1039/D1PY00233C

Chapter 5
Novel Experiment Design to Investigate High-Rate Shear Fracture of Composites

Elias Gerstein, Tyler Robertson, Andrew Baumgardner, Paul Custodio, and Nathan Spulak

Abstract Fiber reinforced composites are extremely desirable materials for aerostructure applications due to their inherent high strength to weight ratios. For safe application of these materials in such structures, it is important to understand their mechanical behavior during high-rate loading. High-rate shear testing of fiber reinforced composites provides important data for understanding the material's high-rate mechanical response. However, there is no generally accepted standardized method for performing high-rate shear tests on fiber reinforced composites. Existing high-rate shear methods typically result in mixed mode loading, rather than deformation and fracture under pure shear conditions. Therefore, a novel unnotched high-rate shear test method for composite materials is investigated. This test method utilizes specialized fixtures to induce direct shear loading on an unnotched test coupon using a split-Hopkinson pressure bar and special high-rate shear fixtures to secure the unnotched test coupon. The novel experiment is designed and simulated using finite element analysis with Ansys LS-DYNA, and preliminary validation tests are performed. The simulation and test results demonstrate a promising concentration of pure shear deformation on the fiber reinforced composite material.

Keywords Fiber reinforced composite · High-rate testing · Shear · Finite element analysis

Introduction

Due to the high strength-to-weight ratios of fiber reinforced composites (FRC's), they have become a desirable material of choice for structural components in the automotive, aviation, and defense industries. The applications FRC's are used for commonly result in them being subject to complex loading conditions across a wide variety of strain rates, such as during vehicle crashes. To safely apply these materials to such conditions, it is therefore important to understand the high strain rate shear behavior of FRC's. However, the current ASTM standard for determining shear properties of composite materials is designed for low rate loading, and does not produce pure shear loading when adapted for use high strain-rate testing (on the order of $10^3 s^{-1}$) [1]. Furthermore, existing non-standardized tests tend to produce mixed mode loading rather than pure shear loading, or they require complex specimen geometries that that can result in damage during the machining process [2]. Therefore, a novel unnotched shear test method is proposed. This method utilizes a simple and easily machinable unnotched specimen and custom fixtures to directly induce pure shear loading at high-strain rates. The high-rate test method utilizes a tensile split-Hopkinson pressure bar (SHPB) to induce the high-rate shear loading desired. The feasibility of this test is initially investigated via finite element analysis (FEA) simulations using Ansys LS-DYNA. Preliminary validation testing for comparison is also performed at low loading rates utilizing a hydraulic load frame, and demonstrate that pure shear deformation is successfully induced on the specimen. This indicates the unnotched shear test is a feasible method for performing pure shear loading at high-strain rates in future testing. Successful production of high-rate shear stress vs strain data obtained from the designed test method can be implemented into the deformation sub-model of LS-DYNA material model MAT_213 to further increase accuracy of FRC material simulations [4].

Elias Gerstein · Tyler Robertson · Andrew Baumgardner · Paul Custodio · Nathan Spulak
Department of Mechanical and Aerospace Engineering, University of Alabama in Huntsville, Huntsville, AL 35899
e-mail: ejg0017@uah.edu; tr0085@uah.edu; apb0015@uah.edu; pc0051@uah.edu; ncs0023@uah.edu

Karen Kasza et al. (eds.), *Mechanics of Biological Systems and Materials and the Mechanics of Composite, Hybrid & Multifunctional Materials, Vol. 3*, Conference Proceedings of the Society for Experimental Mechanics Series, https://doi.org/10.13052/9788743808299-5

Experimental Design and Methods

The proposed test procedure for achieving pure shear loading utilizes a specially designed slotted fixture to hold an unnotched FRC shear specimen. The unnotched shear specimen, shown in Figure 1 (a) is secured within both sides of the slotted test fixture, as illustrated in Figure 1 (b). Then, the grip sections of the fixture are secured within the specimen grips of the hydraulic load frame.

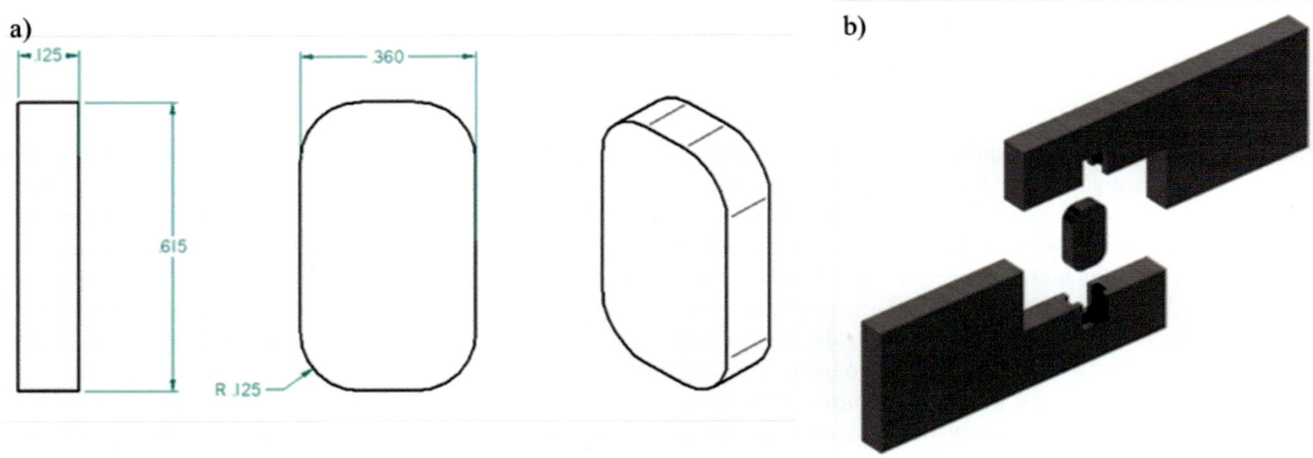

Fig. 1 (a) Unnotched FRC shear specimen (dimensions in inches), and (b) schematic of slotted shear test fixtures with unnotched specimen

Initial FEA simulations of the proposed test setup were done using LS DYNA. One fixture was modeled as a fully constrained rigid material while the other fixture was modeled as rigid material with a prescribed motion. The specimen itself is modeled as a Toray T700S carbon fiber and G83-CM prepreg laminate [2] using the LS DYNA orthotropic elastic material model, MAT_002 [5]. The specimen was then modeled as a solid mesh with a mesh size of 0.305 mm. The purpose of this simulation is to demonstrate the test setup's ability to produce pure shear loading in the test specimen; therefore, the simple orthotropic elastic material model suffices.

Initial low-rate tests are performed on a hydraulic load frame to assess the feasibility and accuracy of the designed test technique. The experiments are performed at the UAH Experimental Mechanics and Multiphysics Modeling (E3M) lab. The specimen material used for the low-rate tests is a polymeric-carbon unidirectional laminate in a quasi-isotropic layup. Low-rate tests are carried out using the fixture from Figure 1 and a hydraulic load frame. To allow for three-dimensional digital image correlation (DIC) analysis, images are taken using two 5 MP cameras, and the specimen and fixtures are covered in a black and white spray paint speckle pattern. The low-rate test is run until fracture of the specimen is observed.

Experimental Results

A comparison of the simulated shear strain and the experimentally observed shear strain prior to specimen fracture is shown in Figure 2. Additionally, the simulated shear strain across along the vertical centerline of the specimen is graphed at various stages as the deformation progresses in Figure 3. The simulation shows the specimen experiences clear shear strain concentration at the center, and this is verified in the preliminary test DIC data which aligns well with the design simulation response. The strain data along the centerline shown in Figure 3 illustrates that the specimen has a highly localized shear strain concentration in the center, with the strain rapidly decreasing towards the outer edges of the specimen.

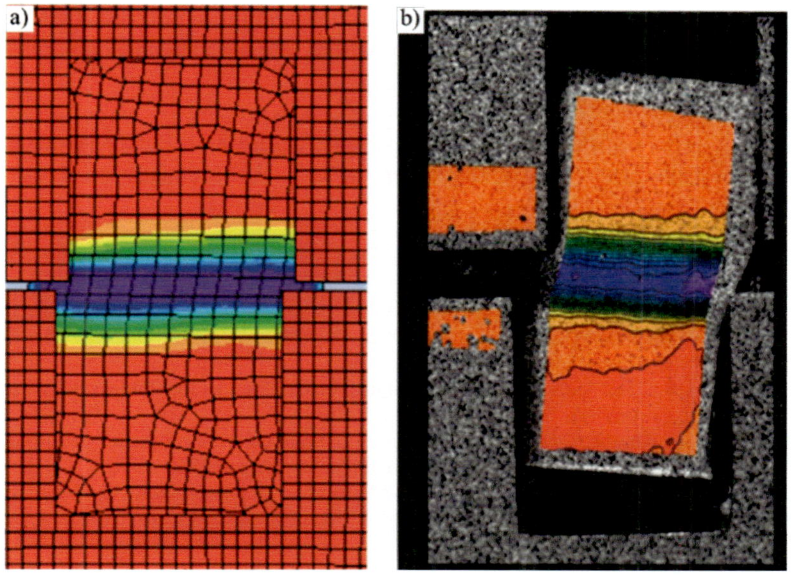

Fig. 2 Shear strain concentration of novel unnotched shear test: (a) LS-DYNA simulation and (b) DIC data from preliminary low rate test

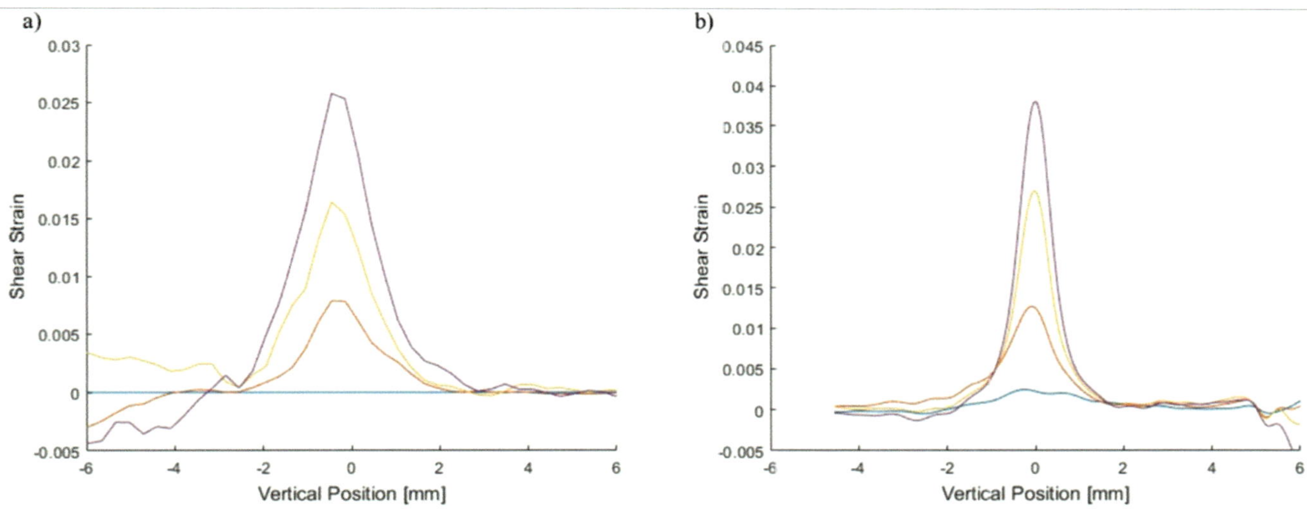

Fig. 3 Shear strain at various stages of deformation along specimen vertical centerline for (a) LS-DYNA simulation and (b) preliminary low rate test

Conclusion

Simulations and low-rate shear testing of a composite material are performed to provide validation of the proposed shear test method. The purpose of the test method is to induce pure shear loading on an FRC specimen. Initial simulations performed using LS-DYNA show the feasibility of this test method, as an obvious concentrated shear strain band is developed within the specimen over the course of the simulation. Low-rate tests are also completed to further provide validation. Through DIC analysis of the low-rate tests, the shear strains are determined over the surface of the FRC specimen. Once again, an obvious concentrated shear band is developed in the specimen, and the shear distribution found is very similar to what was determined in the LS-DYNA simulation. Furthermore, shear strain waterfall plots developed from both the simulation and test match very closely in shape and progression. Thus, the shear concentration proceeds through the course of the test as desired. Future work for this test method will involve carrying out high-rate split-Hopkinson bar testing and will utilize other material systems.

Acknowledgments This research was performed with support from the University of Alabama in Huntsville's Research and Creative Experiences for Undergraduates (RCEU) program and the NASA EPSCoR Rapid Response Research (R3) program award No. 80NSSC24M0151

References

1. D30 Committee. (n.d.). *Test Method for Shear Properties of Composite Materials by the V-Notched Beam Method.* ASTM International. https://doi.org/10.1520/D5379_D5379M-12
2. Yang, P. (2023). Development of Experimental Techniques and Constitutive material model for Unidirectional carbon fiber reinforced polymer (Doctoral dissertation, The Ohio State University).
3. Weng, F., Fang, Y., Ren, M., Sun, J., & Feng, L. (2021). Effect of high strain rate on shear properties of carbon fiber reinforced composites. Composites Science and Technology, 203, 108599. https://doi.org/10.1016/j.compscitech.2020.108599
4. Hoffarth, C., Rajan, S. D., Goldberg, R. K., Revilock, D., Carney, K. S., DuBois, P., & Blankenhorn, G. (2016). Implementation and validation of a three-dimensional plasticity-based deformation model for orthotropic composites. *Composites Part A: Applied Science and Manufacturing*, *91*, 336-350.
5. "LS-DYNA® KEYWORD USER'S MANUAL." (2021). Livermore Software Technology Corporation.

Chapter 6
Experimental and Numerical Analysis of Dynamic Behavior of Cork Material

Veronica Ilari, Carlo Sabbatini, and Marco Sasso

Abstract In this research, the agglomerated cork material is considered for energy absorption purposes. Indeed, the stress-strain curve of cork describes the distinctive mechanical behavior of cellular materials: it can withstand large deformations under compression loading with a smooth transition from elastic to plateau region, and then from plateau to densification stage. In order to characterize the visco-hyperelastic behavior, compression tests have been conducted at different strain rates on cork samples. Low and intermediate strain rate tests have been carried out by means of standard electromechanical and pneumatic testing machines, whereas high strain rates have been carried out using a split Hopkinson pressure bar. The compression tests have been extended including the unloading stage, where the load is progressively removed; in this way, it was also possible to describe the damage evolution within the material according to the Mullins effect theory, which has been borrowed here from the literature on rubber-like materials. It was implicitly assumed that the same behavior occurs in dynamic conditions. The energy absorption capability of a shock absorber made of cork can be evaluated by means of puncture tests that are typically conducted using drop tower rigs. Such tests have been numerically reproduced in Abaqus/Explicit software, where the main features of the material behavior have been included by developing user defined subroutines. The punch deceleration peak and rebound speed, together with the residual deformation and damage of the cork absorber, have been analyzed as performance parameters.

Keywords Cork · Strain rate · Energy absorption · Hyperelasticity · Viscoelasticity

Introduction

In the last decades, cellular materials have been used in various engineering applications, particularly where high energy absorption capability is required. These materials are also valued for their damping and isolation properties, making them essential in impact protection systems. As a result, they are commonly employed in crash test devices designed for personal safety (such as in road accident, where these materials are used for the production of motorcycle helmets) and for protection of things/products (such as in packaging) [1] [2].

Cork is a natural material that is obtained from the outer bark of the cork oak (*Quercus suber*). It has a cellular structure with small closed cells (4-20 x 107 cels/cm^3) that form a highly ordered honeycomb pattern, a characteristic that stems from its biological formation process. Cork shows an interesting set of properties, including low values of density, permeability and energy transfer coefficients with a remarkable elastic behavior and high physical, chemical and biological stability [3].

When subjected to compression, cellular materials like cork can suffer a large deformation, maintaining relatively constant stress levels until densification occurs. This behavior allows them to absorb large amounts of energy. The stress-strain curve of cork exhibits three distinct phases: an initial linear elastic phase at low stresses, primarily governed by the bending of cell walls; a long plateau region, where the material undergoes progressive collapse while dissipating energy through structural deformation; and a densification phase, in which the stress rises sharply due to the complete compaction of the cellular structure [4] [5].

Veronica Ilari · Carlo Sabbatini · Marco Sasso
Dipartimento di Ingegneria Industriale e Scienze Matematiche
Università Politecnica delle Marche, 60131, Ancona, Italy
e-mail: m.sasso@univpm.it; c.sabbatini@pm.univpm.it; v.ilari@pm.univpm.it

Karen Kasza et al. (eds.), *Mechanics of Biological Systems and Materials and the Mechanics of Composite, Hybrid & Multifunctional Materials, Vol. 3*, Conference Proceedings of the Society for Experimental Mechanics Series, https //doi.org/10.13052/9788743808299-6

Beyond its energy absorption capability, cork also possesses a remarkable recovery capacity. This means that its deformation is mostly elastic, allowing it to return to its original shape once the load is removed. This characteristic is particularly advantageous, because it allows the material to maintain its energy absorbing capacity almost unchanged, even after repeated impacts [6] [7].

To characterize the visco-hyperelastic behavior of cork, quasi-static and dynamic compression tests were carry out with a strain rate between 10^{-3} and 10^{3} s^{-1}. The resulting stress-strain curves were used to calibrate the constitutive model parameters that describe cork's mechanical response. In particular, the hyperelastic behavior was modeled by the Ogden hyperfoam formulation; while viscoelasticity was described through a Prony series model (generalized Maxwell model). Additionally, the unloading phase was analyzed to evaluate damage evolution of the material and the potential release of stored elastic energy that can occurre after the impact. The Mullis effect was incorporated to describe the unloading phase [8].

To validate the proposed constitutive model, numerical simulations were performed using the finite element Abaqus/Explicit software. The experimental compression tests were reproduced in the simulation environment, integrating the material properties through user-defined subroutine (VUMAT).

Finally, the energy absorption performance of a cork based shock absorber was evaluated through puncture tests, conducted using a drop tower. These tests were numerically reproduced in Abaqus/Explicit software. The comparison between numerical and experimental results serves as a crucial validation step, ensuring the model's accuracy in predicting the mechanical response of cork under different loading conditions.

Materials and Methods

Compression tests were performed on agglomerate cork samples (Fig.1a) with a base size of 12 x 12 mm and a height of 15 mm. They have a density of 140 kg/m^3. Tests with a strain rate of 10^{-3}, 10^{-1} and 10^{1} s^{-1} were conducted through a servo-pneumatic machine model Siplan® (Fig.1b), equipped with a 3 kN load cell and capable of reaching a piston speed of 100 mm/s, while dynamic tests with a 10^{3} s^{-1} strain rate were performed with a Split Hopkinson Pressure bar (SHPB).

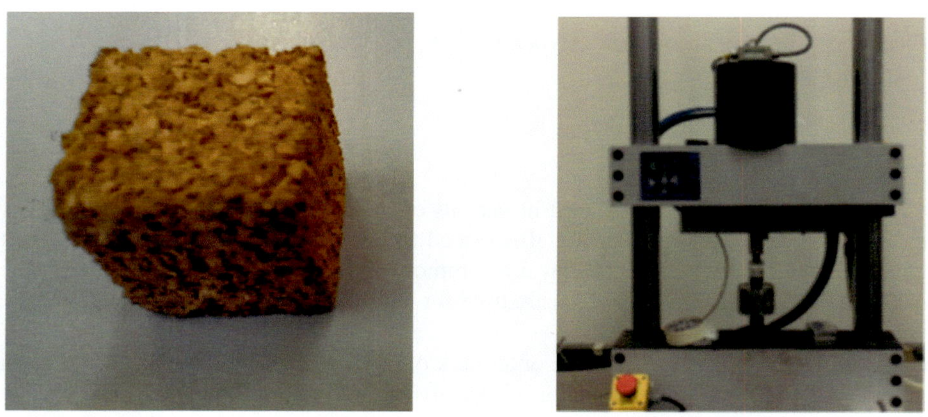

(a) (b)

Fig. 1 a) Agglomerated cork sample; b) Servo-pneumatic testing machine

For the tests conducted at 10^{-3} and 10^{-1} s^{-1}, the samples were placed on a flat plate connected to the load cell, located on the fixed crosshead of the machine. The piston was then moved with precision, until it touches the upper surface of the sample, marking the start of the test.

For the tests conducted ad 10^{1} s^{-1}, however, the piston was initially raised to a sufficient distance from the sample, allowing it to complete the acceleration phase from zero to a 100 mm/s velocity. This strategy allowed to keep the deformation of the samples as uniform as possible, compatible with the capabilities of the machine. To analyze the relaxation phase of the material in tests with 10^{-3} and 10^{-1} s^{-1} strain rate, the deformation was manteined unchanged at the end of the loading ramp for a period equal to the duration of the ramp itself. Subsequently, the load was progressively reduced by moving the piston backwards until the applied load is completely removed.

Dynamic test at 10^{3} s^{-1} strain rate were performed using a Split Hopkinson Pressure Bar (SHPB), shown in Fig.2. The calibration and the fitting of this system for the study of soft materials are described in [9] [10]. Furthermore, the SHPB

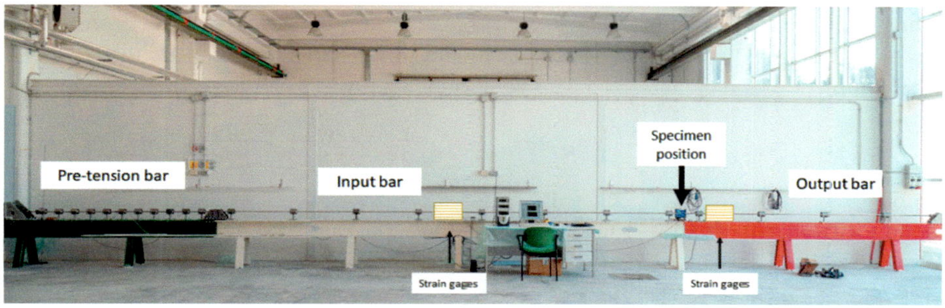

Fig. 2 Split Hopkinson Pressure Bar installed at Marche Polytechnic University

configuration allowed a partial valutation of the relaxation of the sample, extending the analysis of reflected and transmitted waves. This made it possible to partially measure the unloading phase after reaching the maximum load, without a holding phase [11].

Material Modelling

Constitutive models

The constitutive models that describe the material were defined using a VUMAT, that is a user defined subroutine implemented in Abaqus/Explicit.

In this analysis, the agglomerated cork is mainly modeled as a hyperelastic material, using the Ogden hyperfoam model. It is a non linear model, usually used to characterize elastomeric foams that exhibit hyperelastic behavior. It is also intended for finite strain applications where the material can deform elastically to large strains, up to 90 % strain in compression. This model is based on a strain energy potential W. This model is implemented as a built-in material within the Abaqus FE commercial code as "Hyperfoam". The strain energy function of order N is defined:

$$W = \sum_{1=1}^{N} \frac{2\mu_i}{\alpha_i^2} \left[\lambda_1^{\alpha_i} + \lambda_2^{\alpha_i} + \lambda_3^{\alpha_i} - 3 + \frac{1}{\beta_i} (J^{-\alpha_i \beta_i}) \right] \tag{1}$$

where μ_i, α_i and β_i are material constants, $\lambda_{1,2,3}$ are the stretch ratios in the three principal directions, and J is the volume ratio, $J = \lambda_1 \lambda_2 \lambda_3$. The principal stresses are computed taking the partial derivative of Eq.(1) with respect to λ. The experiments also highlighted that cork has a Poisson's ratio of almost zero, thus making the use of a compressible hyperelastic model more appropriate. The μ_i coefficient is directly correlated to the initial shear modulus G, while β_i is related to Poisson's ratio, ν_i, by the expressions:

$$G = \sum_{i=1}^{N} \mu_i \qquad \nu_i = \frac{\beta_i}{1 + 2\beta_i} \tag{2}$$

Therefore, the null Poisson's ratio determines that $\beta_i = 0$. In uniaxial case $\lambda_2 = \lambda_3 = 1$, the principal stresses are obtained from:

$$\sigma = \frac{\partial W}{\partial \lambda_1} = \sum_{1=1}^{N} \frac{2\mu_i}{\alpha_i \lambda_1} (\lambda_1^{\alpha_i} - 1) \tag{3}$$

The Eq.(3) is used to fit the best parameters of the model using the quasi-static compression curve [12].

The agglomerate cork shows a viscoelastic behavior that emerges in dynamic load conditions. Consequently, the viscoelastic behavior must also be modeled in conjunction with the hyperelastic model. The Prony series (or generalized Maxwell) model was chosen. This model considers the material as a branch with a spring representing the long-term stiffness G_∞, in parallel with a series of N_G Maxwell branches, where a spring and a viscous damper are present. Each Maxwell branch is characterized by a relaxation time τ_i. The global relaxation modulus of such a system is given by:

$$G(t) = G_0 \left[g_\infty + \sum_{i=1}^{N_G} \left(g_i e^{-t/\tau_i} \right) \right] \tag{4}$$

In Eq.(4), g_i are the relative moduli, so that $g_\infty + \sum g_i = 1$; G_0 represents the short term, or instantaneous, stiffness that is obtained as the sum of all branches' stiffness. The long term and short term stiffnesses are related as $G_\infty = g_\infty \cdot G_0$. The stress along time in the i-th layer of the Generalized Maxwell model is given by:

$$(\mathbf{S}_i^{2d})_{n+1} = exp\left(-\frac{\Delta t}{\tau_i^G}\right)(\mathbf{S}_i^{2d})_n + \alpha_i^G exp\left(-\frac{\Delta t}{2\tau_i^G}\right)\left[\frac{d\Phi}{d\mathbf{C}_{n+1}} - \frac{d\Phi}{d\mathbf{C}_n}\right] \tag{5}$$

where $\mathbf{S}_i^{2d}$ is the second Piola-Kirchhoff stress tensor and $\mathbf{C}$ is the right Cauchy-Green deformation tensor. The true stress can be obtained from:

$$\sigma = \frac{2\mathbf{F}\mathbf{S}^{2d}\mathbf{F}^T}{J} \tag{6}$$

where $\mathbf{F}$ is the deformation gradient [13].

The total stress can be obtained from the sum of the hyperleastic stress with the N_G viscoelastic stresses.

In order to correctly model permanent energy dissipation effects and stress softening in agglomerate cork, Mullins effect is implemented, providing an extension to the elastomeric foam model [14]. Therefore, this damage model is used to include the damage present in elastomeric foams, modeling the energy absorption in foam components subjected to dynamic loading, with high strain rates compared to the characteristic relaxation time of the foam. In this model, energy dissipation effects are considered by introducing an augmented strain energy density function of the shape:

$$W(\lambda_i, \eta) = \eta W(\lambda_i) + \phi(\eta) \tag{7}$$

The function $W(\lambda_i, \eta)$ is a continuous function of the damage variable, η, and is related to the damage function $\phi(\eta)$. The damage variable, η, varies continuously during the course of deformation and always satisfies $0 < \eta < 1$, with $\eta = 1$ at the points of the primary curve (described by the Hyperfoam model). Taking into account the Mullins effect, the stresses are calculated by:

$$\sigma(\lambda_i, \eta) = \eta\sigma(\lambda_i) \tag{8}$$

where $\sigma(\lambda_i)$ is the stress corresponding to the primary behavior of the foam at the current strain level λ_i. Then, the stress is obtained by simply scaling the stress of the primary behavior of the foam by the damage variable η. Eq.8 is used when the material is at an energy potential W that is lower than the maximum energy potential W_m experienced by the material itself at the end of the loading phase. The damage variable is generally assumed to be represented by the so called "error function":

$$\eta = 1 - \frac{1}{r} erf\left(\frac{W_m - W}{m + \beta W_m}\right) \tag{9}$$

where r,m and β are material parameters that govern the shape of the unloading curve. Note that $r > 1$, $m \geq 0$, $\beta > 0$ [15].

Validation

In order to obtain the coefficients of the models used to describe the cork behavior, an optimization technique was implemented. Specially, a cost function, which represents the difference between the experimentally obtained stresses and those predicted analytically, is minimized.

To describe the compressible hyperelastic behavior, a fourth-order Hyperfoam model was adopted, which requires the identification of 8 parameters, namely the coefficients μ_i and α_i with $i = 1 \dots 4$.

The viscoelastic behavior was instead modeled by a 10-term Prony series, where the Maxwell branches have relaxation times uniformly distributed on a logarithmic scale. Consequently, the optimization problem focuses on the determination of the two extreme relaxation times τ_{min} and τ_{max} and the 10 related moduli g_i. Lastly, Mullins effect requires the calibration of three parameters r,m and β. The optimization process was implemented in Matlab. All the optimized parameters are listed in Table 1, 2 and 3 for the Ogden hyperfoam model, Prony series model and Mullins damage, respectively.

Table 1 4th order Ogden hyperfoam parameters

μ_1	α_1	μ_2	α_2	μ_3	α_3	μ_4	α_4	$\beta_{1 \dots 4}$
4.236	7.532	-5.579	5.580	2.543	4.146	0.007	-3.137	0

After material model calibration, the impact tests were simulated by means of the commercial Finite Element software. In particular, the Abaqus/Explicit code was adopted because it is well suited for implementing user-defined constitutive models by means of user-defined Fortran subroutines (VUMAT).

Table 2 Prony series with 10 terms parameters

τ_{min}	τ_{max}	g_1	g_2	g_3	g_4	g_5	g_6	g_7	g_8	g_9	g_{10}	g_∞
1E-5.608	1E1.322	0.140	0.703	0.046	0.023	0.012	0.006	0.007	0.009	0.006	0.005	0.042

Table 3 Mullins damage parameters

r	m	β
1.067	0.740	0.628

Compression tests were performed at different strain rates on cork block, in order to reproduce as faithfully as possible the experimental tests. The cork is modeled as a 2D model costrained to the bottom side. The plate that load the cork is discretized as a rigid surface and a constant velocity equal to the experimental values. The strain rate analyzed are 10^{-3}, 10^{-1}, 10^{1}, and 10^{3} s^{-1}. The results of both the experimental tests and the analytical modeling, expressed in terms of stress-strain curves, are shown in Fig.3a and in Fig.3b. The blue curves represent the experimental points, while the curves in red represents the total stress predicted by the constitutive model.

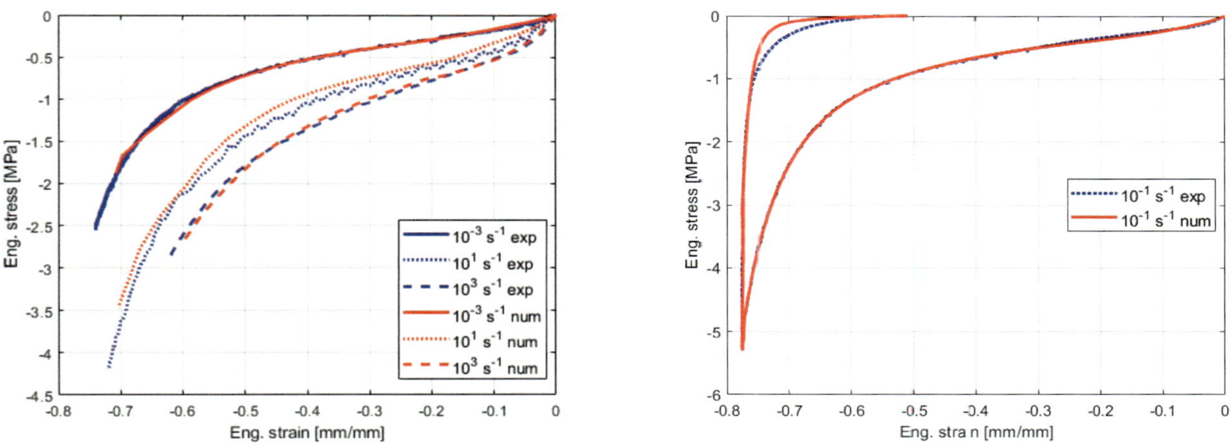

(a) (b)

Fig. 3 a) Compression stress strain curves at 10^{-3}, 10^{1} and 10^{3} s^{-1} strain rate; b) Compression stress strain curves at 10^{-1} s^{-1} strain rate

The Fig.3a presents the results for the compression tests at 10^{-3}, 10^{1} and 10^{3} s^{-1}. The stress-strain curve at the lowest strain rate validates the hyperelastic model, while the two curves corresponding to the highest strain rates confirm the accuracy of the viscoelastic model.

The Fig.3b illustrates the validation of the complete model, including damage, which is captured through the Mullins effect. This test was performed at a strain rate of 10^{-1} s^{-1}.

Test Case

In the previous sections, we defined the constitutive models that best describe the behavior of agglomerated cork. We then validated the model through the simulation of an impact test at different strain rates.

In this section, we present a test case: puncture tests performed using a hemispherical rigid indenter. Puncture tests were performed on cylidrical slabs of cork by means of a drop tower machine. This kind of test is very popular as it provides immediate information on the energy absorption. However, here it has been used for verifying the material models calibration. The specimen is placed between two supports with a 40 mm hole, and is deformed by a falling punch having a hemispherical impact head of 12.7 mm diameter. The speed of the punch is measured by an optical sensor. This test serves as an additional validation step, assessing the model's ability to predict material response under localized loading conditions.

The tests were performed at two different velocities: 0.79 m/s and 1.27 m/s.

The puncture tests have been replicated numerically by means of the FE model and material subroutine VUMAT. The finite element model was developed to closely replicate the experimental setup, ensuring consistency in boundary conditions and loading conditions. A 2D axisymmetric model of the cork block and the impacted area, along with the impactor, was created. The cork block is meshed with 2D axisymmetric reduced integration elements (CAX4R) [16]. The impactor is discretized as a rigid shell, with a 3.055 kg mass and an initial velocity equal to the experimental values. The cork was laterally constrained, as in the experimental test.

Material properties obtained from the previous calibration were assigned to the numerical model, incorporating both hyperelastic and viscoelastic behavior, as well as Mullins damage effects.

The results of the finite element simulations for the puncture tests at impact velocities of 0.79 m/s and 1.27 m/s are presented in Fig.4 and Fig.5, respectively. These contour maps illustrate the distribution of logarithmic strain at three key stages of the impact event: (a) the initial contact between the indenter and the cork specimen, (b) the moment of maximum penetration, and (c) after rebound, when partial recovery of the deformation occurs.

At the instant of maximum penetration (Fig.4b and Fig.5b), it is evident that the deformation in the case of the higher impact velocity (1.27 m/s) is significantly greater than in the lower velocity case (0.79 m/s). The strain distribution in Fig.5b shows a more concentrated and intense deformation region beneath the indenter, indicating a stronger material response to the higher loading rate. This suggests that at increased impact velocities, the cork undergoes greater compression and energy absorption before rebound.

Fig.4c and 5c shows the partial recovery of the strain that occurs when the impact rebounds. It can be seen that the indentation left by the impact is less pronounced in Fig.4c than in Fig.5c and the logarithmic strain is reduced.

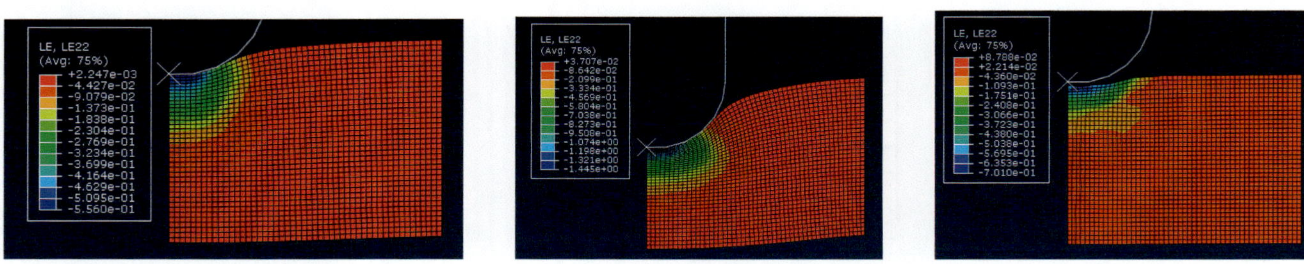

(a) (b) (c)

Fig. 4 Contour maps of logarithmic strain from FE simulations of impact tests at 0.79 m/s a) at the beginning; b) at the instant of maximum penetration; c) after rebound

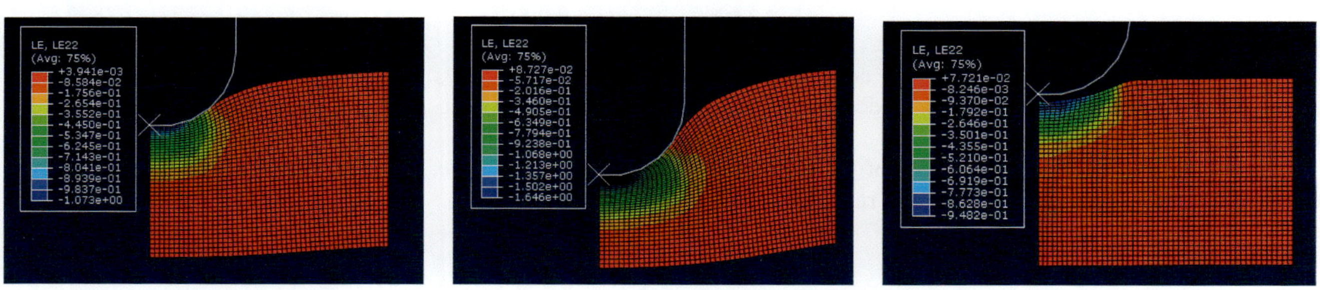

(a) (b) (c)

Fig. 5 Contour maps of logarithmic strain from FE simulations of impact tests at 1.27 m/s a) at the beginning; b) at the instant of maximum penetration; c) after rebound

Conclusion

In this work, we have presented the mechanical characterization and constitutive modelling of agglomerated cork subjected to compression loading. The compression stress-strain curves exhibit the typical behavior of cellular materials. Constitutive models incorporating hyperelasticity, viscoelasticity and the Mullins damage effect were developed and calibrated. These models were then validated through numerical simulations using a VUMAT subroutine, which aimed to accurately reproduce compression tests at different strain rates and provide a comprehensive description of the material's dynamic behavior.

To further assess the constitutive model under more complex loading conditions, we conducted a puncture test at two different impact velocities: 0.79 m/s and 1.27 m/s. The experimental and numerical approaches were integrated to validate the material models, ensuring they accurately capture the complex response of cork under impact loading.

These results confirm that the proposed material models are capable of accurately predicting the cork's behavior under dynamic loading conditions. This validation makes the models a valuable tool for future research on agglomerated cork and similar materials, offering insights into the material's performance in energy absorbing applications and impact resistant designs.

The successful validation of the cork material model through these puncture tests not only reinforces its relevance for practical applications but also provides a solid foundation for future studies.

References

1. Fernandes, F.A.O., Sousa, R.J.A.D., and Pascoal, R.J.S. "Modelling impact response of agglomerated cork". *Materials & Design*, 58:499–507 (2014)
2. Utzeri, M., Cebeci, H., and Kumar, S. "Autonomous sensing architected materials". *Advanced Functional Materials*, 2411975 (2024)
3. Pereira, H. "The rationale behind cork properties: A review of structure and chemistry". *BioResources*, 10 (2015)
4. Anjos, O., Pereira, H., and Rosa, M.E. "Tensile properties of cork in axial stress and influence of porosity, density, quality and radial position in the plank". *Eur. J. Wood Prod.*, 69:85–91 (2011)
5. Gibson, L.J. and Ashby, M.F. *Cellular Solids: Structure and Properties. 2nd ed.* Cambridge University Press (1997)
6. Rumianek, P., Dobosz, T., Nowak, R., Dziewit, P., and Aromiński, A. "Static mechanical properties of expanded polypropylene crushable foam". *Materials*, 14 (2021)
7. Sasso, M., Lattanzi, A., Farotti, E., Sarasini, F., Sergi, C., Tirillò, J., and Mancini, E. "Constitutive modelling of the dynamic behavior of cork material", Challenges in Mechanics of Time Dependent Materials, Mechanics of Biological Systems and Materials & Micro-and Nanomechanics, Volume 2. Conference Proceedings of the Society for Experimental Mechanics Series. Springer, Cham. https://doi.org/10.1007/978-3-030-86737-9_8, pages 57–63 (2022)
8. Mancini, E., Utzeri, M., Farotti, E., Lattanzi, A., and Sasso, M. "Dlp printed 3d gyroid structure: Mechanical response at meso and macro scale". *Mechanics of Materials*, 192:104970 (2024)
9. Sasso, M., Antonelli, M.G., Mancini, E., Radoni, M., and Amodio, D. "Experimental and numerical characterization of a polymeric hopkinson bar by dtma". *International Journal of Impact Engineering*, 103:50–63 (2017)
10. Martarelli, M., Mancini, E., Lonzi, B., and Sasso, M. "On the sensors calibration of polymeric hopkinson bars for dynamic testing of soft materials". *Measurement Science and Technology*, 29 (2018)
11. Utzeri, M., Sasso, M., Deshpande, V.S., and Kumar, S. "Multiscale experiments and predictive modeling for failure mitigation in additive manufacturing of lattices". *Advanced Materials Technologies*, 9 (24):2400457 (2024)
12. Kossa, A. and Berezvai, S. "Novel strategy for the hyperelastic parameter fitting procedure of polymer foam materials". *Polymer Testing*, 53:149–155 (2016)
13. *ANSYS Mechanical APDL Theory Reference.*
14. Ptak, M., Kaczynski, P., Fernandes, F., and de Sousa, R.A. "Assessing impact velocity and temperature effects on crashworthiness properties of cork material". *International Journal of Impact Engineering*, 106:238–248 (2017)
15. Zhang, M.G., Xu, W., Wu, T., Zhang, X.D., Zhang, H., Li, Z., Zhang, C., Jiang, H., and Chen, M. "Investigation on mullins effect of rubber materials by spherical indentation method". *Forces in Mechanics*, 4:100037 (2021)
16. Sasso, M., , Sarasini, F., Mancini, E., Lattanzi, A., Tirillò, J., and Sergi, C. "Experimental characterization and numerical modelling of the impact behavior of pvc foams". *Experimental Mechanics*, 63:823-837 (2023)

Chapter 7
Unexpected Time Dependence in Nanofibrillated Cellulose Gels

Samir Patel and Jacob Notbohm

Abstract Nanofibrillated cellulose (NFC) gels behave similarly to a viscoelastic fluid, as the fibers create an entangled network interacting through van der Waals forces. NFC gels show complex mechanical behaviors, such as shear thinning and shear banding. Previous work has observed time-dependence on short time scales, observing relaxation times on the order of a few minutes, which is consistent with poroelastic behavior. Here, we report the surprising observation of time-dependent behavior over time scales of a few hours, implying that another mechanism dictates this behavior. NFC gels were thoroughly mixed with fluorescent particles for imaging on a spinning-disk confocal microscope. The gels were imaged for multiple hours, and the strains were computed using Fast Iterative Digital Image Correlation. Experiments first quantified strains in gels subjected to compression and in unloaded gels with various aspect ratios. The gels deformed for a few hours in the loading device regardless of whether a load was applied, and gels with high aspect ratios deformed over a few hours to become more circular. These results suggest that NFC gels may exhibit capillary behavior, which could be the mechanism for time dependence at long time scales. Further experiments quantified displacements in NFC gels when water was introduced by contracting microspheres to gather insight on poroelastic behavior. The gels moved inward towards the expelled water at short time scales, even though they were not adhered to the contracting microspheres. This work motivates future work to study in detail the effects of water on the mechanics of NFC gels over a wide range of time scales.

Keywords Hydrogels · Cellulose · Poroelasticity · Capillary · Time-dependence

Introduction

Nanofibrillated cellulose (NFC) gels are biodegradable and renewable, making them appealing for many uses, including 3D printing, food packaging, and drilling fluid [1]. Cellulose gels behave similarly to a viscoelastic fluid, as the fibers create an entangled network with van der Waals interactions. The different production methods used to create NFC can change the rheology and mechanics of the system by altering fiber geometry and cross-linking [2, 3]. NFC suspensions show complex fluid-like behaviors, such as shear thinning in a rheometer [1, 4]. Quantifying the rheology of NFC gels using commercial rheometers is challenging, because the measured modulus depends on sample thickness [5]. This is likely due to shear banding and strain localization near the boundaries of the sample, which cause inconsistencies in measured data of the rheological properties such as viscosity and shear modulus [6, 7]. These complications make the rheology of NFC gels challenging to quantify.

Cellulose gels are hydrophilic and swell substantially in the presence of water. Swelling behavior has been studied at larger length scales in cellulose sponges [8, 9] with results showing that the sponges swell with water on time scales of a few minutes. Previous work has measured time-dependent behavior in cellulose gels to be on the order of minutes in stress relaxation experiments [10]. Prior work has also suggested that cellulose gels exhibit poroelasticity, with a compressive relaxation time of around 16-160 s for cellulose gels of diameter on the order of centimeters [11]. Poroelasticity is common in other materials as well, including biological tissues [12], the cytoplasm within cells [13], and tendons [14]. The time scale of tens to hundreds of seconds for relaxation of centimeter-sized samples is also common. For example,

Samir Patel · Jacob Notbohm
Department of Mechanical Engineering
e-mail: spatel222@wisc.edu; jacob.notbohm@wisc.edu

Samir Patel · Jacob Notbohm
Engineering Mechanics Program, University of Wisconsin-Madison, 1500 Engineering Dr, Madison, WI 53706
e-mail: spatel222@wisc.edu; jacob.notbohm@wisc.edu

Karen Kasza et al. (eds.), *Mechanics of Biological Systems and Materials and the Mechanics of Composite, Hybrid & Multifunctional Materials, Vol. 3*, Conference Proceedings of the Society for Experimental Mechanics Series, https://doi.org/10.13052/9788743808299-7

polyvinyl alcohol-polyacrylic acid hydrogels of size on the order of centimeters have a stress relaxation time on the order of minutes [15].

In this report, we experimentally quantify the time-dependent properties of cellulose gels, observing surprising time dependence occurring at time scales on the order of hours. Three experiments were conducted to study the mechanics of the NFC gels. (i) Strains were quantified after NFC gels were subjected to compression. The mean strains over time showed a time dependence on the order of hours, which led to further experiments to explain this phenomenon. It was found that this long time dependence was due to interactions between the NFC gel and boundaries, suggesting possible capillary effects. (ii) The deformation of gels of various aspect ratios was discerned by quantifying strains when placed on hydrophilic substrates for a few hours, with results showing the gels deformed to become circular, suggesting a strong effect of surface tension. (iii) To study NFC gels' interaction with water at length scales close to the size of fibers, water was introduced to the cellulose gels by embedding poly(N-isopropylacylamide) (PNIPAAm) microspheres, which contract when heated [16, 17, 18]. Upon contraction, the microspheres left a void space filled with water, which the NFC gels expanded into over time scales of tens of minutes.

Methods

Tempo-Oxidized Nanofibrillated Cellulose (T-NFC) was provided by the Forest Products Laboratory (Madison, WI) as a 1.11 weight percent concentration. The NFC gels were vortex mixed with Alexa Fluor carboxylated fluorescent particles of 0.5 μm diameter and excitation and emission wavelengths of 580 and 605 nm, respectively (F8812, Thermo Fisher Scientific). The concentration of fluorescent particles in the NFC gels was chosen to be around 3.6 mg per 1 mL NFC. We compared pipette mixing and vortex mixing. Vortex mixing the fluorescent particles and NFC gel provided a more consistent mixture of fluorescent particles in the gel. Gels embedded with fluorescent particles were used in all experiments.

PNIPAAm microspheres were created through an adapted version of the methods from ref. [18]. The microspheres were created using an oil-water emulsion. Cyclohexane, span-80 (S0060, TCI America), N-isopropylacrylamide (415324, Millipore Sigma), bis-acrylamide (1610142, Bio-Rad), ammonium persulfate (1610700, Bio-Rad), and tris-buffered saline were mixed on a stir plate in a nitrogen atmosphere. TEMED was added during the mixing and the reaction occurred overnight. The microspheres were then washed with ethanol, deionized water, and 1× phosphate buffered saline.

In all experiments, NFC gels embedded with fluorescent particles were imaged on a spinning-disk confocal microscope (Yokogawa CSU-X1) with 50 μm pinholes, a 20× 0.95 NA water-immersion objective (Nikon), and a Zyla sCMOS camera (Andor). Image specifications were set through IQ3 acquisition software (Andor). All images were collected with a 150 ms exposure time. All z-stacks had a step size of 0.325 μm. The number of z-steps collected varied from about 50 (for experiments applying uniform strain) to about 150 (for experiments with PNIPAAm microspheres). When imaging NFC gels without PNIPAAm microspheres, the z-planes chosen for imaging were at least 60 μm above the glass substrate to avoid significant surface effects. When imaging gels with PNIPAAm microspheres, we chose planes at or close to the microspheres' centers. In both experiments, we experienced some drifting in all directions, which we manually accounted for during image acquisition. After acquiring the images, the z-stacks from different time points were registered to each other, enabling selection of the same focal plane at all time points for subsequent analysis by digital image correlation (DIC).

A custom-built loading device was used to apply compression to the NFC gels. The applied strain was actuated by a micrometer. The device was designed to fit in the microscope stage, so that images of the gels before and after applying strain could be taken. Strains were quantified in NFC gels when subjected to compression over a few hours. Strains were also quantified in unloaded gels placed either inside or outside of the loading device. To further test capillary behavior, NFC gels embedded with fluorescent particles were pipetted onto glass slides such that they either had a higher length (0.75 in) than their width (0.25 in) or were circular (radius 0.25 in). We imaged these gels every 5 min for up to 3 hr for subsequent analysis by DIC. To gain insight on poroelastic behavior, we carefully pipette-mixed PNIPAAm microspheres into NFC gels embedded with fluorescent particles, so the microspheres would not be damaged before the experiment. The microspheres acted as void inclusions in the gel, which decreased in volume when the temperature was raised from the reference temperature (29°C) to 39°C [17]. The temperature was controlled using an H301 stage top incubator (Okolab). This was mounted directly on the microscope stage so that imaging and changing the temperature of the environment could be performed simultaneously. During contraction, the PNIPAAm microspheres became hydrophobic and expelled water. We imaged the gel before and after the contraction of the microspheres, which typically occurred within 20-30 min. Microspheres selected for imaging were at least a field of view away from other microspheres to avoid effects from other microspheres.

DIC was used to quantify the displacements and strains, enabling us to avoid errors caused by strain localization. Fast Iterative DIC [19] was performed on all image sets to quantify the in-plane displacement fields. We used a subset size of

64 × 64 pixels (20.8 × 20.8 μm^2) and a spacing of 16 pixels (5.2 μm). DIC was performed by comparing each image to the corresponding image in the reference (undeformed) configuration. The strains were calculated from the gradient of displacements. Numerical derivatives were taken by finding slopes in the x and y directions in a kernel of a 5×5 window of points, which generates the displacement gradient at the center of the window [20]. This process was repeated throughout the entire image to compute the strain fields.

Results

We first quantified the DIC noise floor by quantifying strains from two consecutive images of an undeformed NFC gel embedded with fluorescent particles imaged with the confocal microscope (Fig. 1a). From the images, in-plane displacements were computed by DIC, and the in-plane components of the strain tensor were computed from the displacement field. The strain fields and histograms of ε_{xx}, ε_{yy}, and ε_{xy} did not exceed 0.5% (Fig. 1b,c). The histograms show that the noise was random and resembled a Gaussian distribution. The standard deviation of noise from this analysis was approximately 0.1% strain.

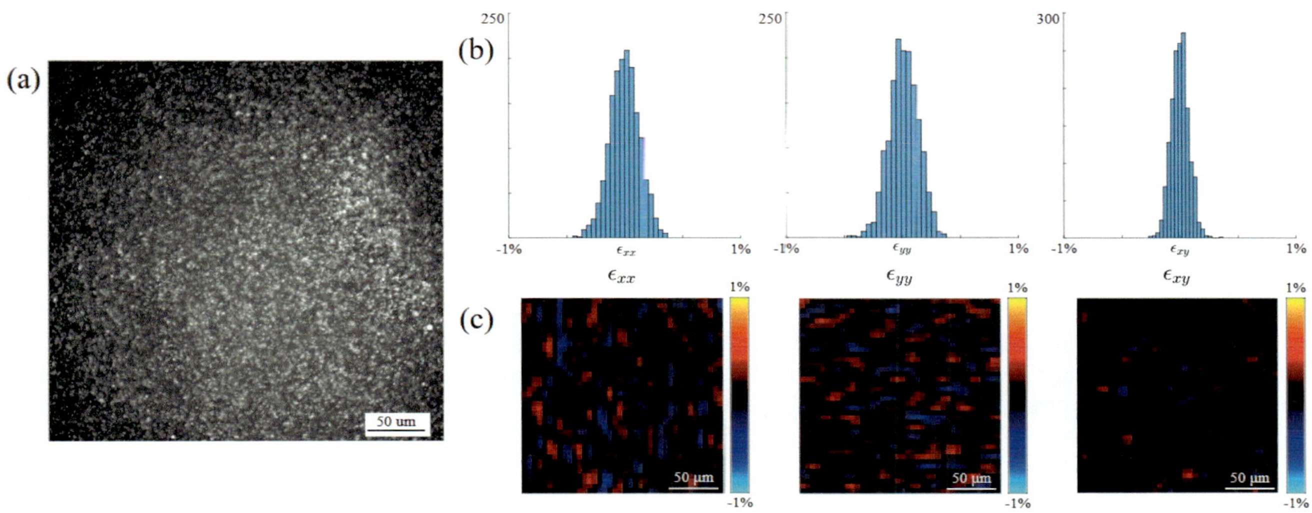

Fig. 1 Noise floor analysis on an NFC gel embedded with fluorescent particles. (a) Representative confocal image of NFC gel embedded with fluorescent particles. (b) Histograms of strains. (c) Strain fields. Results from panels b and c show that the noise in measured strains was less than 0.5%.

Next, we applied compression to NFC gels, as shown in Fig. 2a, using a custom loading device that was mounted on the microscope stage. Images were collected in a region of interest near the center of each gel, followed by calculation of displacements and strains by DIC. When the gels were compressed in the x direction, there was an immediate response in the strain in the region of interest, showing $\varepsilon_{xx} < 0$ which is expected for compression. The variability in strain over space, as indicated by the standard deviation, was relatively small, being approximately 1%, indicating that NFC gels are relatively homogeneous, even at scales of tens of microns, which differs from other fibrous materials such as gels of collagen [17, 18, 21]. Interestingly, ε_{xx} changed over time, increasing over a period of a few hours, which is unexpected given that the loading device applied a constant displacement at the boundaries of the sample. For the strains in the region of interest to change over time, the gel must have undergone a complex time-dependent deformation with increasing strain in the region of interest balanced by decreasing strain in other locations. The time dependence was also surprising, as it occurred on time scales of hours. Prior work on compressive relaxation on cellulose gels computed a relaxation time of around 16-160 s for samples that were centimeters in size [11]. In the theory of poroelasticity, the poroelastic time scale τ is related to sample size L such that the expression τ/L^2 is constant [22]. Using this expression and the prior data τ = 160 s and L = 2 cm, we would expect our samples having size of $\approx$ 2.5 cm would have a characteristic poroelastic time of around 4 min, which is far shorter than the time scales for deformation shown in Fig. 2a. Hence, there must be some other mechanism that drives the deformation of these gels.

NFC gels were next placed in the loading device but not subjected to compression, with the expectation that the gel would remain undeformed. Surprisingly, ε_{xx} increased over a time period of a couple of hours (Fig. 2b). This may suggest that the deformation was caused by interaction between the sample and the loading device. To verify this, we placed gels on a glass coverslip outside of the loading device. When quantifying strains with DIC, ε_{xx} was near zero for all time points (Fig. 2c). This indicates that the sample interacted with the loading device, causing time-dependent deformation. We reasoned that the interaction could be through differences in surface tension of the NFC gel in the different conditions. This suggests that NFC gels may exhibit elasto- or visco-capillary behavior.

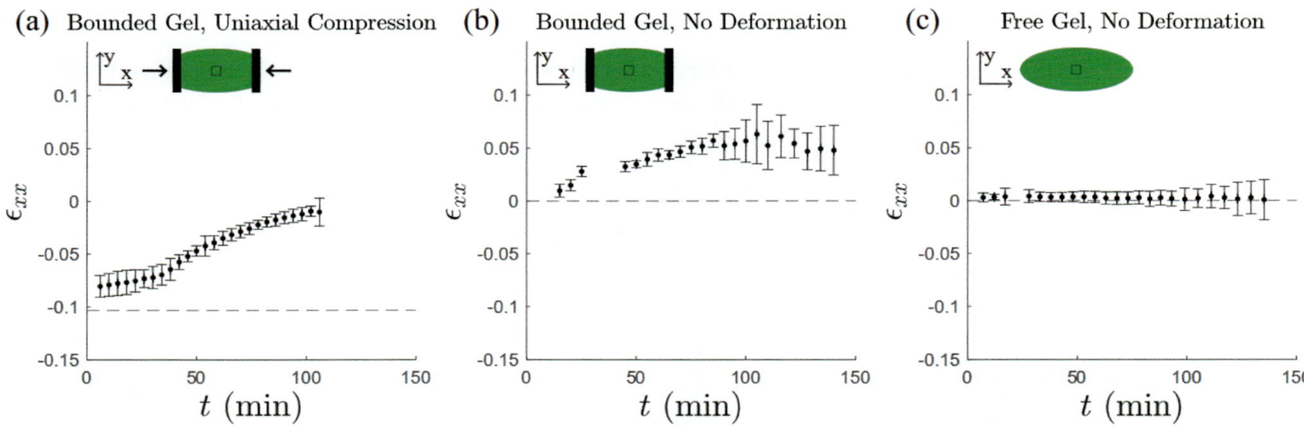

Fig. 2 Strains in experiments with NFC gels. (a) NFC gels were placed in the loading device and were subjected to uniaxial compression in the x direction. Time $t = 0$ denotes when the deformation was applied. (b) NFC gels were placed in the loading device but not subjected to any loading. ε_{xx} shows that despite no loading, the strain increased over a period of a couple of hours. (c) NFC gels were not placed in the loading device. The strains in these gels were near zero for the duration of the experiment. All images show cartoons of NFC gels (green) and the location imaged by the microscope (squares). Panels a and b show the gel in the loading device (black). In all images, ε_{xx} is quantified through DIC and averaged at each time point. Dots in each image denote the mean strain, and error bars denote standard deviations over space. The dashed line denotes the macroscale strain applied by the loading device on the gel.

As a simple initial quantification of capillary behavior, we placed NFC gels of various aspect ratios on glass coverslips and imaged them for a few hours. Some gels had length-to-width ratios of 3:1, while other gels were circular. The average strains are shown in Fig. 3. For the elongated gels, the normal strains along the short axis (y direction) increased over an hour (Fig. 3b). This indicates that the gels deformed to become more circular. Deformations of the gels were apparently permanent, as the gels did not return to their initial size and shape. By contrast, the circular gels remained essentially undeformed with strains being approximately zero for all time points (Fig. 3c). Because we did not apply a load on the gels during these experiments and elongated gels became circular, the gels' deformation was likely caused by surface tension, consistent with capillary behavior.

We next quantified the response of NFC gels to introduction of a void space containing water. This experiment was intended to give a sense of poroelastic swelling at a length scale close to that of the fibers, which is a smaller length scale than the experiments in Figs. 2–3. To create void space in the NFC gels, we used microspheres of PNIPAAm, a hydrogel that contracts and expels water when heated. Microspheres were added to NFC gels embedded with fluorescent particles. An image was collected at 29°C followed by increasing the temperature to 39°C, which caused the microspheres to contract and expel water. A second image was collected for subsequent quantification of displacements by DIC. Fig. 4a shows a schematic of the described experiment, wherein a PNIPAAm microsphere (black) contracts and pushes water out, indicated by the white region. For most engineering materials, this experiment would cause no deformations, because the PNIPAAm microspheres were not attached to the surrounding material, but the NFC gel could potentially swell in response to the free water. Fig. 4b displays a representative image in the reference state, with the PNIPAAm microsphere being the black circle in the center of the image and the contrast provided by the fluorescent particles embedded in the NFC gel. After contraction, the fluorescent particles appeared to move inward (Fig. 4c), indicating inward deformation of the NFC gel. The inward deformation was confirmed in the displacements quantified by DIC, with a maximum magnitude of displacement of $\approx$5 μm (Fig. 4d). This inward motion was likely due to swelling of the NFC into the region containing water expelled by the contracting microsphere.

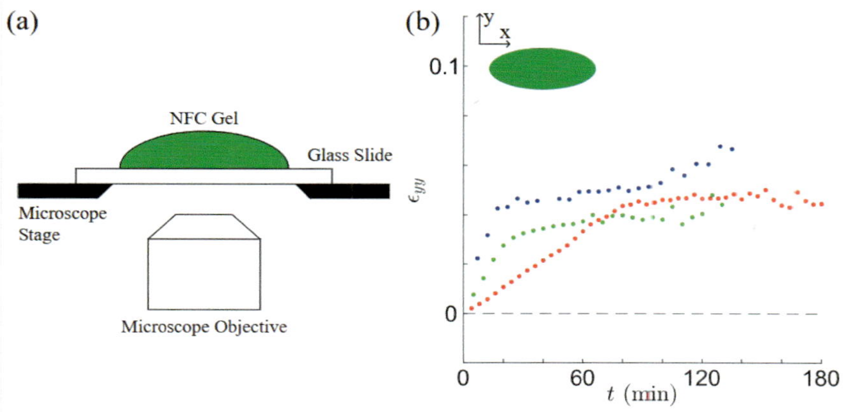

Fig. 3 Mean strains in free NFC gels of two different aspect ratios. (a) Imaging setup for quantifying strains in elongated or circular gels. (b) Elongated gels, where the length of the gel ($\approx$0.75 in) was much larger than its width ($\approx$0.25 in). Here, ε_{yy} increased over time, indicating elongation of short axis so the gel became more circular. (c) Circular gels, with radii close to the width of the long gels. Here, $\varepsilon_{yy} \approx 0$ for all time points. In subfigures (b-c), colors denote results from different samples. In all experiments between both shapes, ε_{xx} was approximately zero for all time points.

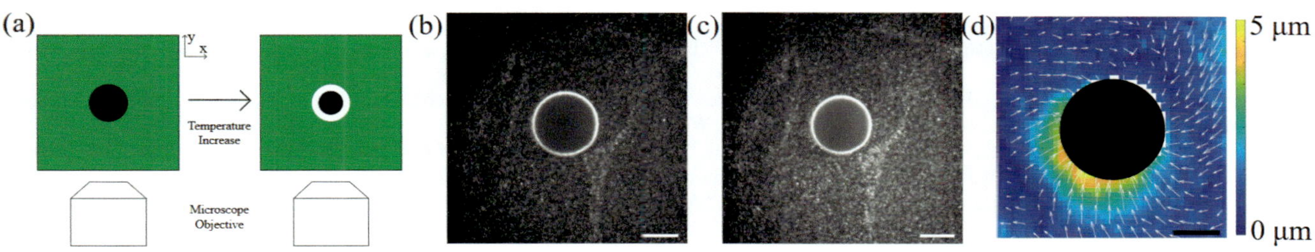

Fig. 4 Local displacements in an NFC gel. (a) Schematic of contracting PNIPAAm microspheres in NFC gel. The microspheres are not physically or chemically adhered to the NFC. The white space around the contracted microsphere denotes water that NFC can expand into. (b-c) Images of NFC gel before (b, 29°C) and after (c, 39°C) contraction of a representative PNIPAAm microsphere. (d) Local displacements quantified through DIC. The color indicates magnitude of displacements, and arrows show direction and magnitude. White regions in the displacement field are erroneous regions that are removed from the analysis as in ref. [17]. The black circle indicates the location of the microsphere; it is larger than the size of the microsphere, as DIC can not get proper data at boundaries in the image. All scale bars denote 50 µm.

Conclusions

In this report we have shown an unexpected time dependence, with deformations occurring on the order of hours in gels made of NFC. Subsequent investigation showed that elongated gels on hydrophilic substrates deformed to become more circular, suggesting capillary behavior may be the underlying cause for the long time dependence. The experiments with PNIPAAm microspheres revealed that NFC gels swelled into the water expelled by the PNIPAAm. These observations, combined with prior studies on poroelastic effects in NFC gels, indicate that the mechanics of NFC gels results from a combination of swelling, poroelasticity, and capillary behavior.

Future work will design experiments to test this combination of properties in more detail. A challenge to consider in future work is drying of the gels, which occurs on time scales of hours, similar to the time scales of deformation observed in this study. In the experiments shown in Fig. 2, it would not be possible to wait for the sample to equilibrate in the loading device before testing, because the hours of equilibration time would cause the sample to dry excessively. Experiments are being designed to use samples having dimensions that minimize the surface area exposed to air, which will enable rigorous quantification of mechanical properties.

Acknowledgments We thank John Considine and the Forest Products Laboratory for providing the NFC used in this study. This work was supported by the Department of Mechanical Engineering at UW-Madison and by the United States Endowment for Forestry and Communities grant number 22-CA-11111129-049. The United States Endowment for Forestry and Communities, Inc. is a not-for-profit corporation that collaborates with partners in the public and private sectors to advance systemic, transformative, and sustainable change for the health and vitality of the nation's working forests and forest-reliant communities.

References

1. M.-C. Li, Q. Wu, R. J. Moon, M. A. Hubbe and M. J. Bortner, "Rheological Aspects of Cellulose Nanomaterials: Governing Factors and Emerging Applications," Advanced Materials, vol. 33, p. 2006052, 2021. W409W10080
2. K. Missoum, M. N. Belgacem and J. Bras, "Nanofibrillated Cellulose Surface Modification: a Review," Materials, vol. 6, p. 1745–1766, 2013. W409W10080
3. H. A. Khalil, Y. Davoudpour, M. N. Islam, A. Mustapha, K. Sudesh, R. Dungani and M. Jawaid, "Production and Modification of Nanofibrillated Cellulose Using Various Mechanical Processes: a Review," Carbohydrate Polymers, vol. 99, p. 649–665, 2014. W409W10080
4. S. Mohan, G. H. Koenderink and K. P. Velikov, "Inelastic Behaviour of Cellulose Microfibril Networks," Soft Matter, vol. 14, p. 6828–6834, 2018. W409W10080
5. A. Pakzad, "Size Effects on the Nanomechanical Properties of Cellulose I Nanocrystals," Journal of Materials Research, vol. 27, pp. 528-536, 2011. W409W10080
6. O. Nechyporchuk, M. N. Belgacem and F. Pignon, "Current Progress in Rheology of Cellulose Nanofibril Suspensions," Biomacromolecules, vol. 17, p. 2311–2320, 2016. W409W10080
7. O. Nechyporchuk, M. N. Belgacem and F. Pignon, "Rheological Properties of Micro-/Nanofibrillated Cellulose Suspensions: Wall-Slip and Shear Banding Phenomena," Carbohydrate Polymers, vol. 112, p. 432–439, 2014. W409W10080
8. J. Ha, J. Kim, Y. Jung, G. Yun, D.-N. Kim and H.-Y. Kim, "Poro-Elasto-Capillary Wicking of Cellulose Sponges," Science Advances, vol. 4, p. eaao7051, 2018. W409W10080
9. J. Kim, J. Ha and H.-Y. Kim, "Capillary Rise of Non-Aqueous Liquids in Cellulose Sponges," Journal of Fluid Mechanics, vol. 818, p. R2, 2017. W409W10080
10. P. Lopez-Sanchez, M. Rincon, D. Wang, S. Brulhart, J. R. Stokes and M. J. Gidley, "Micromechanics and Poroelasticity of Hydrated Cellulose Networks," Biomacromolecules, vol. 15, p. 2274–2284, 2014. W409W10080
11. M. R. Bonilla, P. Lopez-Sanchez, M. J. Gidley and J. R. Stokes, "Micromechanical Model of Biphasic Biomaterials With Internal Adhesion: Application to Nanocellulose Hydrogel Composites," Acta Biomaterialia, vol. 29, p. 149–160, 2016. W409W10080
12. E. E. Konofagou, T. P. Harrigan, J. Ophir and T. A. Krouskop, "Poroelastography: Imaging the Poroelastic Properties of Tissues," Ultrasound in Medicine & Biology, vol. 27, p. 1387–1397, 2001. W409W10080
13. E. Moeendarbary, L. Valon, M. Fritzsche, A. R. Harris, D. A. Moulding, A. J. Thrasher, E. Stride, L. Mahadevan and G. T. Charras, "The Cytoplasm of Living Cells Behaves as a Poroelastic Material," Nature Materials, vol. 12, p. 253–261, 2013. W409W10080
14. B. K. Connizzo and A. J. Grodzinsky, "Tendon Exhibits Complex Poroelastic Behavior at the Nanoscale as Revealed by High-Frequency AFM-Based Rheology," Journal of Biomechanics, vol. 54, p. 11–18, 2017. W409W10080
15. P. Chiarelli, D. De Rossi and P. Basser, "Hydrogel Stress-Relaxation," Journal of Intelligent Material Systems and Structures, vol. 4, p. 176–183, 1993. W409W10080
16. M. Proestaki, A. Ogren, B. Burkel and J. Notbohm, "Modulus of Fibrous Collagen at the Length Scale of a Cell," Experimental Mechanics, vol. 59, p. 1323–1334, 2019. W409W10080
17. B. Burkel and J. Notbohm, "Mechanical Response of Collagen Networks to Nonuniform Microscale Loads," Soft Matter, vol. 13, p. 5749–5758, 2017. W409W10080
18. B. Burkel, M. Proestaki, S. Tyznik and J. Notbohm, "Heterogeneity and Nonaffinity of Cell-Induced Matrix Displacements," Physical Review E, vol. 98, p. 052410, 2018. W409W10080
19. E. Bar-Kochba, J. Toyjanova, E. Andrews, K.-S. Kim and C. Franck, "A Fast Iterative Digital Volume Correlation Algorithm for Large Deformations," Experimental Mechanics, vol. 55, p. 261–274, 2015. W409W10080
20. H. Farid and E. P. Simoncelli, "Differentiation of Discrete Multidimensional Signals," IEEE Transactions on Image Processing, vol. 13, p. 496–508, 2004. W409W10080
21. M. Sarkar and J. Notbohm, "Quantification of Errors in Applying DIC to Fiber Networks Imaged by Confocal Microscopy," Experimental Mechanics, vol. 62, p. 1175–1189, 2022. W409W10080
22. A. H.-D. Cheng, Introduction, in "Poroelasticity," S. M. Hassanizadeh and J. Bear, Eds., Theory and Applications of Transport in Porous Media (pp. 1-59), Springer, 2016.W409W10080

Chapter 8
Biofilm Adhesion on TiAl6V4 using Laser Spallation Technique

Sahar Afshari and Martha E. Grady

Abstract Biomedical implants such as orthopedic, dental, cardiac devices, and catheters are prone to infections. The primary factor of implant infections is the adhesion of bacteria to the implant surface, leading to colonization and biofilm formation, which could eventually cause implant failure. Thus, quantifying the adhesion strength of biofilms on the implant surfaces is crucial for characterizing and improving surfaces and materials, and preventing biofilm infections. In this study, the laser spallation technique is utilized to measure the adhesion of biofilm to the implant surface. The aim of this study is to evaluate the adhesion strength of *Staphylococcus aureus* biofilms on Ti6Al4V (Ti64) substrates and to qualitatively compare with *Streptococcus mutans* on pure Ti. For this purpose, a substrate assembly suitable for measuring biofilm and cell adhesion on bulk materials such as Ti64, and additively manufactured Ti and Ti alloys, is employed. The results demonstrated that a substrate assembly comprising a layer of implant material as the substrate and a waterglass layer as the confining layer can be successfully utilized to assess biofilm failure across varying laser fluences. The ongoing challenges that need to be addressed to achieve effective adhesion measurement of cells and biofilms are also described.

Keywords Laser spallation · Biofilms · Biomedical implants · Titanium alloy · Ti6Al4V · Adhesion

Introduction

Biomedical devices such as catheters, cardiac devices, dental implants, and artificial joints have significantly altered the lives of patients and health care. However, their use has made the body more vulnerable to infections. Biomedical device-associated infections comprise 25.6% of all healthcare-associated infections in the US [1]. Among these infections, implant-associated infections, primarily caused by biofilm formation and colonization, cause a significant challenge due to the difficulties in diagnosis and treatment [2]. Biofilms with their ability to resist antibiotics and firmly attach to implant surfaces, can lead to persistent infections and even implant loss in severe conditions. Several factors, such as the type of implant and its location, prevalent microorganisms, surgical environment, and host immune system influence implant-associated infections [3]. Predominant microorganisms in biomedical implant infections vary according to the type and location of implants. For instance, staphylococci are the most common pathogens in orthopedic implant infections, with *Staphylococcus aureus* (*S. aureus*) and *Staphylococcus epidermidis* accounting for two-thirds of cases [4]. *S. aureus* is also the most frequent pathogen in cardiac-related implant infections such as pacemakers, stents, and left ventricular assist devices [3]. In dental implants, biofilm communities are more complex and contain multiple species such as *Streptococci* and *Actinomyces* which are prominent early colonizers, providing conditions favorable for late colonizers [5]. Generally, multiple species tend to be more challenging and difficult to treat than single species infections [6]. However, in vitro assessments of oral biofilms have also been conducted using single species such as *Streptococcus mutans*, which has proven to be an early colonizer in dental caries [7, 8].

Ti and its alloys are widely used metals in dental implants, knee and hip replacements, screws for fracture fixation, artificial hearts, pacemaker cases, and prosthetic heart valves [9, 10]. Ti-6Al-4V, the most commonly used Ti alloy in high-strength and load-bearing applications [9], is selected as the implant material for this study. This alloy is favored for its good corrosion resistance, biocompatibility, superior strength-to-weight ratio, and robust fatigue resistance [11, 12].

Sahar Afshari · Martha E. Grady
Department of Mechanical and Aerospace Engineering, University of Kentucky, Lexington, KY, USA 40508
email: s.afshari@uky.edu; m.grady@uky.edu

Karen Kasza et al. (eds.), *Mechanics of Biological Systems and Materials and the Mechanics of Composite, Hybrid & Multifunctional Materials, Vol. 3*, Conference Proceedings of the Society for Experimental Mechanics Series, https://doi.org/10.13052/9788743808299-8

The proposed study focuses on utilizing bulk materials such as conventional Ti64 and additively manufactured (AM) Ti64 as the substrate to obtain cell and biofilm adhesion. *S. aureus* on Ti64 substrates and *S. mutans* on pure titanium were evaluated as representatives of biofilm infections in orthopedic and dental implants, respectively. Although conventional Ti64 is utilized in the current study, the substrate assembly procedure is also applicable to AM Ti and AM Ti alloy.

In this work, the laser spallation technique is implemented as a non-contact method that generates high stress levels. This technique allows the precise macroscale quantification of cell and biofilm adhesion on substrates as well as localized impingement and multiple adhesion tests on a single substrate assembly. The laser spallation technique employs a high-energy YAG laser pulse that impinges on a substrate and generates a compressive stress wave. The generated stress wave propagates through the substrate, transmitting to the film and eventually reflecting off from the free sample surface [13]. However, due to the unreflective nature of biological films, in situ measurement of cells is not possible, and accurate calibration experiments on the substrates are required. A typical laser spallation specimen consists of a confining layer, an absorbing layer, a substrate, and a thin film. The laser spallation technique has been successfully adopted to measure the interfacial adhesion of cells and biofilm as a thin film on different substrates in numerous studies [14–16].

Experimental Procedures and Results

Fig. 1 (a) illustrates the experimental setup of the laser spallation technique for the biofilm and cell adhesion measurement. A Nd:YAG laser pulse is controlled using an attenuator and focused in a desired spot size using a focusing lens. The controlled and focused YAG laser pulse is then directed upward toward the substrate assembly using a 45° angled mirror.

A 1 mm thick Ti64 sheet purchased from American Metal Xchange, Inc. was cut into 1-inch by 1-inch square samples. The surface roughness, Ra, of Ti64 samples was measured using a Sensofar S neox optical profilometer. at 0.26 ± 0.02 μm from three samples, with measurements taken in three scanned lengths for each sample. One side was coated with a 12 μm thick layer of waterglass (Ricca Chemical sodium silicate) using the Specialty System G3P-8. The substrate assemblies consisted of Ti64 as the substrate and waterglass as the confining layer, as shown in Fig. 1 (b). The waterglass thickness was measured using a Bruker Dektak XT Surface Profiler. The substrates were then adhered to the bottom of a petri dish containing a 13/16-inch hole using the biocompatible NOA86H adhesive and cured under a Lesco CUREMAX Chamber Lamp (Fig. 1 (b)).

In order to culture bacteria, 1 μL of *S. aureus* (ATCC 6538) was taken from frozen stock using an inoculation loop and transferred to a 15 mL centrifuge tube containing 5 mL of Todd Hewitt Yeast (THY) broth. Bacteria were grown in a bead bath at 37 °C for 24 hours. The following day, prepared substrates were sprayed with 70% ethanol and sterilized under UV light for 30 minutes. The grown bacteria were diluted with THY to achieve an OD of 0.7 ± 0.05, which was measured using a GE Ultrospec 8000 spectrophotometer. The substrate was placed in a second petri dish, wrapped with parafilm to protect the waterglass layer from the heat in the incubator, and 1 mL of the 0.7 OD *S. aureus* and THY solution, combined with 3 mL of a 75 mM glucose-THY solution, was added to the substrate assembly. These substrate assemblies were then cultured at 37 °C with 5% CO2 for 24 hours. The media were then removed from the dishes, and the biofilms on the substrates were gently rinsed with phosphate-buffered saline (PBS) to remove any non-colonized bacteria. *S. mutans* (Wild type Xc) was also cultured using the same protocol, except on the second day, a 75 mM sucrose-THY solution was used.

Fig. 1 (c) illustrates *S. aureus* before and after loading at a high iso-fluence experiment of 96 mJ/mm^2 on a Ti64 sample, where the loaded area demonstrated complete biofilm detachment. The substrate assemblies for the *S. aureus* experiment consisted of a 1 mm Ti64 substrate and a 12 μm thick waterglass confining layer. As depicted in Fig. 1 (c), the substrate assemblies successfully led to full spallation at the highest fluence, indicating that the Ti64 substrate was able to absorb the YAG laser energy and generate high compressive stress through the substrate, delaminating the biofilm from the substrates without the need for any additional absorbing layer. However, the thickness of the waterglass layer plays a significant role in the magnitude of the stress wave generated by the YAG laser, with an increase in waterglass thickness significantly increasing the amplitude of the generated stress wave.

The substrate assemblies for *S. mutans* consisted of 9 μm thick waterglass as the confining layer, 100 nm Al absorbing layer, 1 mm thick glass substrate, and 100 nm pure Ti film. *S. mutans* were grown on the Ti surface to mimic a dental implant surface. The grown *S. mutans* were subjected to varying laser fluences which resulted in biofilm ejection at the loaded areas. Figure 1 (d) demonstrates the *S. mutans* biofilm grown on smooth titanium before and after loading. Figure 1 (e) represents the fluorescence microscopy images of the loaded *S. mutans* on smooth titanium. The applied fluence increased from 37.5 to 75.7 mJ/mm^2, leading to a significant increase in the ejection area at higher fluences compared to lower ones. At the highest laser fluence of 75.7 mJ/mm^2, complete ejection of *S. mutans* was observed from the pure Ti substrates across all loaded areas.

Results revealed that the adoption of the substrate assembly for bulk materials such as Ti64 and AM Ti64 has successfully generated a high compressive stress wave for cell and biofilm adhesion measurement studies. This effective implementation of the substrate assembly was achieved by applying a controlled confining layer on the Ti64 substrate to generate sufficient interface stress, compared to the common multilayer substrate assembly that includes a confining layer, an absorbing layer, a substrate, and a thin film. Both the loaded *S. mutans* and *S. aureus* on different substrates exhibited a similar failure area. However, to accurately measure the interfacial adhesion of *S. aureus* on Ti64, further experiments under varying laser fluences should be conducted.

In the preliminary applications of the laser spallation technique for the biological cell adhesion measurement, studies quantified the interfacial adhesion strength of cells and biofilms on substrates by identifying the onset of detachment as the critical threshold, where the minimum laser fluence required to detach cells from the substrate is determined as the critical laser fluence [15, 17, 18]. More recently, the 50% adhesion strength has been reported as the interfacial adhesion strength of the cells and biofilms by researchers [16, 19]. Because of the non-uniform nature of the biological film, instead of a particular laser fluence, a range of fluences is contributed to the onset of failure adhesion [19]. Thus, this 50% adhesion failure represents the adhesion failure of cells and biofilm more effectively. Although the appearance of a dark spallation area has been identified as the failed test in other studies [16, 19], similar to the failed biofilm depicted in Fig. 1 (d), results indicated that the appearance of the bare implant surface may provide a more accurate determination of failure, which may appear lighter in color depending on the tilt of the sample when viewing, as shown in Fig. 1 (c).

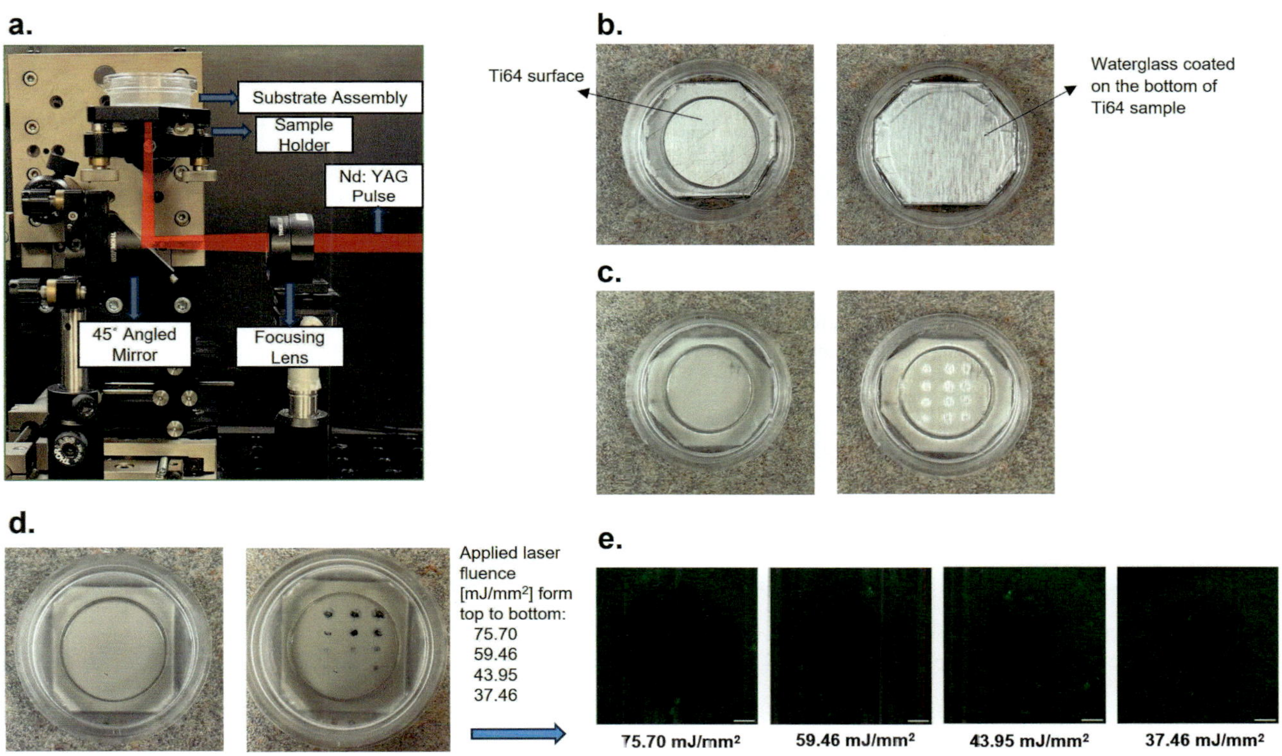

Fig. 1 (**a**) Experimental setup of laser spallation technique for cell and biofilm adhesion measurement. (**b**) Cell and biofilm culture substrate assembly consisted of Ti64 substrate and a waterglass layer as the confining layer. (**c**) *S. aureus* grown on Ti64 before and after loading at an iso-high fluence experiment, substrate assembly consisted of 1 mm Ti64 and 12 μm thick waterglass (**d**) *S. mutans* grown on smooth titanium before and after loading across varying laser fluences. (**d**) Fluorescence microscopy images of spalled biofilms at increasing laser fluence from 37.46 to 75.7 mJ/mm^2, captured using a 4× objective. *S. mutans* biofilm was stained with Syto 9 before loading. The scale bar is 0.5 mm, and the applied laser spot size is 2 mm

Challenges

Several challenges have emerged while progressing towards the goals. Initially, substrates were attached to the bottom of a Petri dish using Dowsil 732 silicone sealant, which caused a chemical reaction with the waterglass layer, resulting in a waterglass dissolution. To address the waterglass dissolution issue, a biocompatible UV-curing adhesive, NOA86H, was applied to uniformly attach the substrate to the Petri dish, which preserved the integrity of the waterglass layer for further experiments as well as reducing the thickness of the interlayer optical adhesive compared to the silicone sealant. One current issue is the degradation of the waterglass layer in the incubator due to exposure to humidity and heat. To preserve the waterglass layer in the incubator, exploring different types of waterglass, varying application methods, or alternative confining layers might be utilized. For calibration experiments, the free surface of the substrates must be reflective enough to produce a distinct fringe pattern and accurately capture the displacement movement. The reflectivity of the Ti64 samples was first examined using a Michelson interferometric setup, where the Ti64 sample failed to generate any reflective fringes. Although hand-polished Ti64 samples showed improved reflectivity, it was still inadequate for capturing in situ data measurements. Consequently, a major ongoing challenge is achieving mirror-like reflectivity in Ti64 for laser interferometric measurements.

Acknowledgments We gratefully acknowledge the support of the National Science Foundation (NSF) through the CAREER Award, Grant No. 2045853. Measurements of waterglass thickness were conducted at the Center for Nanoscale Science and Engineering at the University of Kentucky. We thank Dr. Julius Schoop for providing access to the optical surface profilometer in his lab, which was used to measure the surface roughness of the Ti64 samples. We also thank Dr. Natalia Korotkova for her donation of *S. mutans* bacteria used in this study.

References

1. Magill, S.S., et al., Multistate point-prevalence survey of health care–associated infections. New England Journal of Medicine, 2014. 370(13): p. 1198-1208.
2. Zimmerli, W., A. Trampuz, and P.E. Ochsner, Prosthetic-joint infections. New England journal of medicine, 2004. 351(16): p. 1645-1654.
3. Kandi, V. and S. Vadakedath, Implant-associated infections: A review of the safety of cardiac implants. Cureus, 2020. 12(12).
4. Campoccia, D., L. Montanaro, and C.R. Arciola, The significance of infection related to orthopedic devices and issues of antibiotic resistance. Biomaterials, 2006. 27(11): p. 2331-2339.
5. Belibasakis, G.N., et al., Peri-implant infections of oral biofilm etiology. Biofilm-based Healthcare-associated Infections: Volume I, 2015: p. 69-84.
6. Ebersole, J.L., et al., The periodontal war: microbes and immunity. Periodontology 2000, 2017. 75(1): p. 52-115.
7. Loesche, W.J., Role of Streptococcus mutans in human dental decay. Microbiological reviews, 1986. 50(4): p. 353-380.
8. Daboor, S.M., et al., A review on streptococcus mutans with its diseases dental caries, dental plaque and endocarditis. Indian Journal of Microbiology Research, 2015. 2(2): p. 76-82.
9. Elias, C., et al., Biomedical applications of titanium and its alloys. Jom, 2008. 60: p. 46-49.
10. Liu, X., P.K. Chu, and C. Ding, Surface modification of titanium, titanium alloys, and related materials for biomedical applications. Materials Science and Engineering: R: Reports, 2004. 47(3-4): p. 49-121.
11. Ralls, A., et al., Material design and surface engineering for bio-implants. Jom, 2020. 72(2): p. 684-696.
12. Zhao, S., et al., Compressive and fatigue behavior of functionally graded Ti-6Al-4V meshes fabricated by electron beam melting. Acta Materialia, 2018. 150: p. 1-15.
13. Ehsani, H., et al., Evolution of the laser-induced spallation technique in film adhesion measurement. Applied Mechanics Reviews, 2021. 73(3): p. 030802.
14. Hu, L., et al., Cell adhesion measurement by laser-induced stress waves. Journal of Applied Physics, 2006. 100(8).
15. Shim, J., et al., Measurement of the tensile strength of cell–biomaterial interface using the laser spallation technique. Acta biomaterialia, 2008. 4(6): p. 1657-1668.
16. Boyd, J., N. Korotkova, and M. Grady, Adhesion of biofilms on titanium measured by laser-induced spallation. Experimental mechanics, 2019. 59: p. 1275-1284.
17. Nakamura, H., et al., Glycosaminoglycan degradation reduces mineralized tissue–titanium interfacial strength. Journal of Biomedical Materials Research Part A: An Official Journal of The Society for Biomaterials, The Japanese Society for Biomaterials, and The Australian Society for Biomaterials and the Korean Society for Biomaterials, 2006. 77(3): p. 478-486.
18. Hagerman, E., et al., Evaluation of laser spallation as a technique for measurement of cell adhesion strength. Journal of Biomedical Materials Research Part A: An Official Journal of The Society for Biomaterials, The Japanese Society for Biomaterials, and The Australian Society for Biomaterials and the Korean Society for Biomaterials, 2007. 82(4): p. 852-860.
19. Boyd, J.D., et al., Biofilm and cell adhesion strength on dental implant surfaces via the laser spallation technique. Dental Materials, 2021. 37(1): p. 48-59.

Chapter 9
Towards Single Particle Tracking in Simulated Microgravity-Grown Biofilms

William T. Anderson, Ross Rodriguez, Austin Stallings, Tony Butera, Emmabeth Parrish Vaughn, and Martha E. Grady

Abstract A bacterial biofilm is an adhered collection of bacteria contained in a self-produced extracellular polymeric substance (EPS). The mesh-like EPS provides mechanical and biochemical advantages to a biofilm compared to planktonic cells, enabling the biofilm to better survive on surfaces, and resist neutralization by antibiotics. The growth of antibiotic-resistant biofilms on biological surfaces like bone and tissue, as well as biologically relevant substrates like implants, has significant and negative clinical implications. Additionally, biofilm and microbial growth have been documented on the International Space Station (ISS), including *Staphylococcus aureus,* an opportunistic pathogen that forms biofilms. Past spaceflight studies have indicated that the microgravity environment induces changes to microbial virulence and structure, potentially compromising astronaut safety and health. To simulate the microgravity environment, *S. aureus* biofilms will be grown on a clinostat. Using particle tracking and analysis software, single particle tracking (SPT) techniques will be utilized on polymer-coated quantum dots embedded within the biofilm to measure diffusivity and mesh size. While tests like minimum inhibitory concentration (MIC) are traditionally used when exploring antibiotic resistance, they provide no insight into the mechanism of action of the resistance. Results from this study will provide a quantitative, biophysical perspective into the virulence of biofilms grown in simulated microgravity to inform on antibiotic diffusion, targeted drug delivery, nutrient transport, biofilm modeling, and biofilm mechanics for both space and Earth biofilms. Future work will involve partnering with an implementation partner to develop an automated platform suitable to run the experiment on the ISS in a true microgravity environment.

Keywords Biofilm · Biofilm matrix · Antibiotic resistance · Microgravity · Single particle tracking

Introduction

Bacterial biofilm-associated infections are a major health challenge due to the challenges associated with their eradication. Extracellular polymeric substances (EPS) secreted by bacteria form a matrix around the cells and are characteristic of a biofilm. This matrix gives structural support and protection to the biofilm [1] and affects the penetration and diffusion of antibiotics and nutrients [2]. Measuring antibiotic resistance can be performed with a relatively simple test of minimum inhibitory concentration (MIC) [3]. However, the returned values do little to inform how a biofilm resists antibiotics, which is critical to developing countermeasures. Physically probing a biofilm mesh with nanoparticles and tracking them with single particle tracking (SPT) fluorescence microscopy techniques produces metrics that quantitatively describe how nano-sized particles behave in the biofilm given a dependency on the mesh size and structure. Particle behavior is also relevant to the development of nanocarriers for precision drug delivery.

Along with Earth-based biofilms, space biofilms also pose a potential risk to the continual human presence in space, especially as long-term and deep-space missions continue to be developed, and permanent lunar establishments are expected. Swabbing experiments onboard the International Space Station (ISS) designed to provide a detailed overview of the microbial

William T. Anderson · Ross Rodriguez · Austin Stallings · Tony Butera · Martha E. Grady
Department of Mechanical and Aerospace Engineering, University of Kentucky, Lexington, KY 40508
e-mail: wtan223@uky.edu; rrodriguez34@my.apsu.edu; at.stallings@uky.edu; buteratony123@gmail.com; m.grady@uky.edu

Ross Rodriguez · Emmabeth Parrish Vaughn
Department of Physics, Engineering, and Astronomy, Austin Peay State University, Clarksville, TN 37044
e-mail: rrodriguez34@my.apsu.edu; vaughnep@apsu.edu

Karen Kasza et al. (eds.), *Mechanics of Biological Systems and Materials and the Mechanics of Composite, Hybrid & Multifunctional Materials, Vol. 3*, Conference Proceedings of the Society for Experimental Mechanics Series, https://doi.org/10.13052/9788743808299-9

signature have found that *Staphylococcus aureus*, a biofilm-forming bacterial species, is a common microbe aboard the ISS [4–6]. NASA has recognized a "Risk of Adverse Health Effects Due to Host-Microorganism Interactions" [7] based on past experiments in space, with a suggested high-level research approach that includes "phenotypic evaluation of microorganisms (including virulence studies) and temporal changes for these parameters under both spaceflight and spaceflight analogue conditions" [7]. Biofilm experiments in space (Table 1) are a subset of the microbial experiments that have been completed in space. However, a major focus of the biofilm studies has been genetic analysis, with less emphasis on structural and morphological changes in which SPT techniques excel. Additionally, conclusions drawn from the experiments have been hindered by unreproducible findings [8].

Ground-based simulations of the space environment play an important role in bioastronautics research because of the time and financial limitations associated with in-space research. They may also be required before an on-orbit experiment is approved. On Earth, microgravity can be simulated with temporal dependence by clinostats which continuously rotate biological specimens to balance out gravity's force to near zero. Analogs must also accurately model the diffusion-driven fluidic environment produced by microgravity, meaning platforms must avoid machine-induced stirring/motion of fluids.

To enable single particle tracking in simulated microgravity-grown biofilms, the following materials and techniques are needed: appropriate and well-characterized particles, the ability to accurately track them, and a microgravity simulation platform for biofilms. The focus of this report is a brief review of previous biofilm spaceflight experiments, biofilm SPT experiments and techniques, and current work to merge the two.

Table 1 Previous space biofilm experiments. Adapted and expanded from [9]

<table>
<tr><th>Mission/Launch Vehicle</th><th>Location</th><th>Bacterial Species</th><th>Project Name</th><th>Results Ref.</th></tr>
<tr><td>STS-81</td><td>Space Shuttle</td><td>Burkholderia cepacia</td><td>N/A</td><td>[10]</td></tr>
<tr><td>STS-95</td><td>Space Shuttle</td><td>Pseudomonas aeruginosa</td><td>N/A</td><td>[11]</td></tr>
<tr><td>STS-132</td><td rowspan="2">Space Shuttle</td><td rowspan="2">Pseudomonas aeruginosa</td><td rowspan="2">N/A</td><td rowspan="2">[12]</td></tr>
<tr><td>STS-135</td></tr>
<tr><td>NG-12</td><td>ISS</td><td>Pseudomonas aeruginosa</td><td>Space Biofilms, CU Boulder</td><td>[13]</td></tr>
<tr><td>SpaceX-22</td><td>ISS</td><td>Unknown oral bacteria</td><td>Oral Biofilms in Space, UNLV, Colgate</td><td>N/A—private funding</td></tr>
<tr><td>SpaceX-23</td><td>ISS</td><td>Staphylococcus capitis</td><td rowspan="3">BIOFILMS Experiment, ESA</td><td rowspan="3">N/A as of 02/2025</td></tr>
<tr><td>N/A</td><td>ISS</td><td rowspan="2">Cupriavidus metallidurans OR Acinetobacter radioresistens</td></tr>
<tr><td>SpaceX-27</td><td>ISS</td></tr>
</table>

SPT in Biofilms

Particle tracking techniques have been employed in various biofilm species with several particle types and sizes to investigate diffusive and microrheological properties. As the motion of a particle is influenced by its environment, analysis of this motion through SPT techniques can elucidate the properties of the environment. SPT can broadly be categorized into two categories: one in which particles actively bind to specific targets to track (proteins, molecules, etc.), and one in which particles do not actively bind and are the focus of tracking. The latter is the primary technique for biofilm studies. Fluorescent polystyrene beads ranging from 40-2000 nm are a common particle choice. Larger particles, on the order of the biofilm mesh size, are used for mechanical property measurements like viscoelasticity, while smaller particles are used for diffusion-based metrics—although there is no consensus for a size cutoff. Particles are expected to exhibit confined behavior indicated by a decrease in diffusion coefficients with respect to values in water, and nonlinear mean squared displacements (MSD). Results show that particle size, charge, and coating impact the measured values, with some results displayed in Table 2.

Table 2 Select biofilm SPT studies comparing strain, particle size, and diffusion coefficients

Bacterial Strain	Particle Size (nm)	Diffusion Coefficient ($\mu m^2/s$)	Ref.
Pseudomonas aeruginosa	40	.31493	[14]
	100	.16647	
	100	.03-2.82	[15]
	200	.00061-.00358	[14]
	500	.00165	
Escherichia coli	200	.00137-.034	
	500	.01	[16]
Staphylococcus aureus	40	.11584	[14]
	100	.03549	
	200	.00031-.00074	
	500	.00038	
Burkholderia multivorans	100	.18-3.21	[15]
	200	.11-1.71	

SPT Techniques and Validation

A glycerol-water solution provides a controlled, viscous solution in which a researcher's experimental setup can be confirmed and hydrodynamic radii and control diffusion coefficients can be measured. In a free fluid like a glycerol-water solution, particles undergo Brownian motion. After tracking, MSDs can be used to extract diffusion coefficients. The Stokes-Einstein equation is then employed to relate the diffusion coefficient to the particle's hydrodynamic radius. Note that the most widely used equation relies on the assumption of a spherical particle.

FIESTA, a particle tracking software, has been used to track quantum dots in cells [17] and was utilized to generate particle tracks from videos of quantum dots in a glycerol water solution (Fig. 1 a). @msdanalyzer, a MATLAB package [18], was used for track analysis (Fig. 1 b-c).

Particle videos contain the spatiotemporal data extracted by particle tracking software to create the trajectories for analysis. Thus, the quality of the videos is critical to the generation of trajectories that accurately represent the true motion of the particles. Video frame rate and resolution can be considered the most important parameters and should be maximized to record a high-quality video. However, increasing the frame rate usually comes at the expense of exposure time—consequently decreasing the overall video and particle brightness. Similarly, increasing video resolution usually comes at the expense of frame rate. These parameters must be balanced and depend on the type of particle and camera.

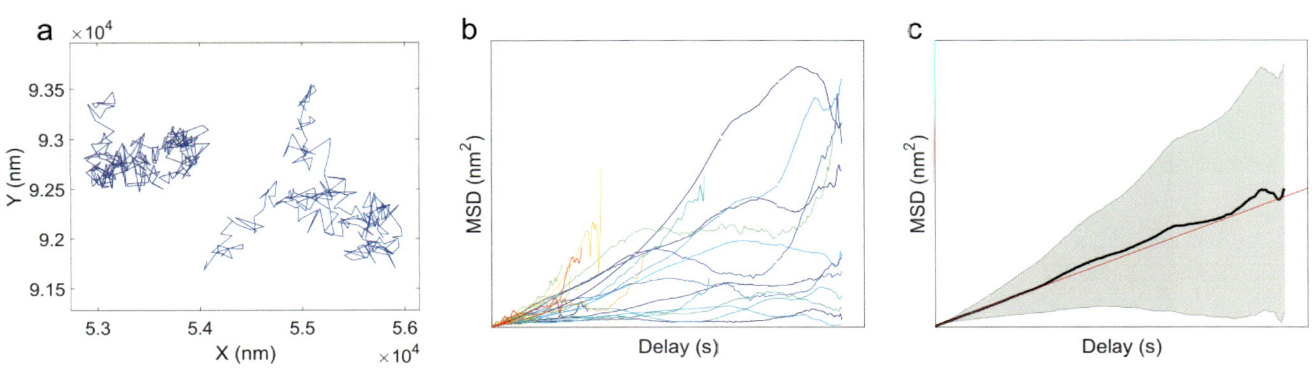

Fig. 1 SPT analysis of Thermo Fisher QDs in 90 w/v% glycerol. Two characteristic particle tracks generated by FIESTA (a) with full population of track MSDs (b) and weighted MSD average (black line) and fitted line (red line) (c)

Accurate Particle Selection

The choice of particle in a particle tracking experiment is important. Accurate analysis of passive, individual probe tracks in a biological material requires careful consideration of what probe to use, as inappropriate probe choice can mask results or lead to wrong conclusions. As noted earlier, several factors including probe size, charge, coating, fluorescence, and concentration each contribute to how a probe will interact with the sample and experimental setup of choice. While commercial probes are easily accessible to researchers who do not have the expertise, facilities, or time to create custom probes or modify existing probes, these probes are often not well characterized by their respective vendors. Consequently, the authors of this study aim to provide an in-depth characterization of 10+ purchasable probes to assist future researchers make informed decisions. Preliminary results with a focus on three quantum dots are presented here.

Quantum dots (QDs) are semiconductor nanocrystals and are desirable for their small size and inherent, stable fluorescence. QDs are often coated with a polymer like polyethylene glycol (PEG) when used in biological applications. This addition can serve several purposes [19]: to protect cells from the toxic heavy metal core, prevent specific binding if the polymer is nonreactive/nonfunctionalized (necessary for passive particle tracking), or to hold reactive/functional groups for specific binding (necessary for active particle tracking).

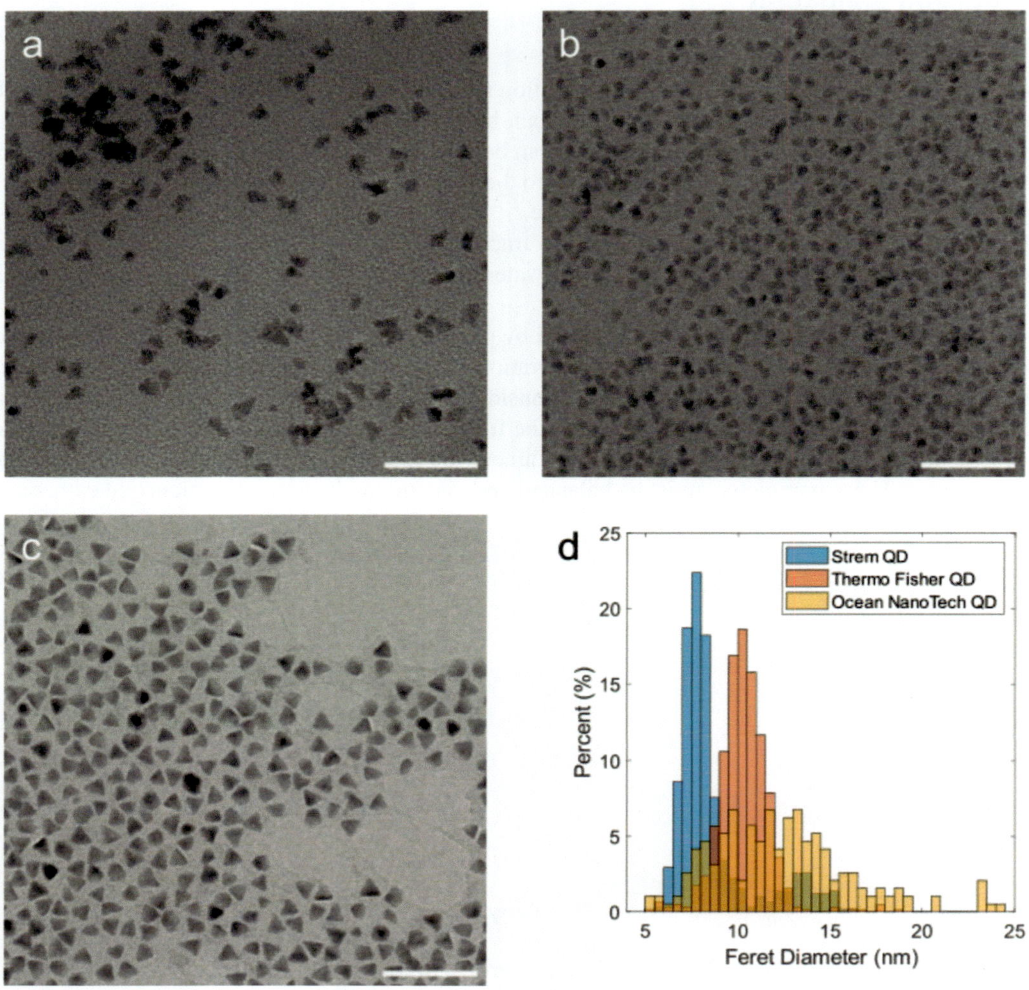

Fig. 2 Representative TEM images of commercially available quantum dots from Ocean NanoTech (a), Strem (b), and Thermo Fisher (c). Scale bars are 50 nm. Feret diameters (d) of Thermo Fisher and Strem QDs display narrow distributions while Ocean NanoTech QDs exhibit a wide distribution of sizes

Materials and Methods

Three commercially available QDs were purchased for characterization: Thermo Fisher Qdot™ 605 ITK™ Amino (PEG) Quantum Dots (Q21501MP), Ocean NanoTech PEG Quantum Dots 620 nm (QMG620-02), and Strem 620 nm CdSe/ZnS core/shell quantum dots with carboxylic acid (48-1640). TEM images of the QDs were taken on a Talos F200X TEM (Fig. 2). The concentration of particles in the images is arbitrary. Diluted samples were prepared on lacey carbon-supported 300 mesh copper TEM grids. QD Feret diameters (maximum caliper length) of representative samples were measured with ImageJ.

Preliminary Size Characterization Results

Thermo Fisher QDs had an average Feret diameter ± standard deviation of 10.3 ± 1.4 nm. Ocean NanoTech QDs had an average Feret diameter ± standard deviation of 12.4 ± 3.8 nm. Strem QDs had an average Feret diameter ± standard deviation of 8.6 ± 2.1 nm. Observational analysis of the TEM images (Fig. 2), confirmed by measured values, shows that the Thermo Fisher QDs are highly uniform in size and shape but are triangular, while the Ocean Nanotech QDs are nonuniform in both size and shape. The Strem QDs are uniform in both size and shape and are the most spherical of the three samples, potentially making them the most desirable for tracking and subsequent analysis.

Acknowledgements and Funding Sources The authors are grateful for the TEM images taken by experts in the University of Kentucky Electron Microscopy Core and to the National Science Foundation Graduate Research Fellowship Award No. 2239063 and National Science Foundation CAREER Award Grant No. 2045853 for providing funding.

References

1. H. C. Flemming, T. R. Neu, and D. J. Wozniak, "The EPS matrix: the "house of biofilm cells"," (in eng), J Bacteriol, vol. 189, no. 22, pp. 7945-7, Nov 2007, doi: 10.1128/jb.00858-07.
2. R. Mirghani et al., "Biofilms: Formation, drug resistance and alternatives to conventional approaches," (in eng), AIMS Microbiol, vol. 8, no. 3, pp. 239-277, 2022, doi: 10.3934/microbiol.2022019.
3. B. Kowalska-Krochmal and R. Dudek-Wicher, "The Minimum Inhibitory Concentration of Antibiotics: Methods, Interpretation, Clinical Relevance," (in eng), Pathogens, vol. 10, no. 2, Feb 4 2021, doi: 10.3390/pathogens10020165.
4. A. Checinska Sielaff et al., "Characterization of the total and viable bacterial and fungal communities associated with the International Space Station surfaces," Microbiome, vol. 7, no. 1, p. 50, 2019/04/08 2019, doi: 10.1186/s40168-019-0666-x.
5. S. Xiao, K. J. Venkateswaran, and S. C. Jiang, "The risk of Staphylococcus skin infection during space travel and mitigation strategies," Microbial Risk Analysis, vol. 11, pp. 23-30, 2019/04/01/ 2019, doi: https://doi.org/10.1016/j.mran.2018.08.001.
6. H. Jang, S. Y. Choi, and R. J. Mitchell, "Staphylococcus aureus Sensitivity to Membrane Disrupting Antibacterials Is Increased under Microgravity," (in eng), Cells, vol. 12, no. 14, Jul 21 2023, doi: 10.3390/cells12141907.
7. "Human Research Program Integrated Research Plan," NASA, 07/2023 2023.
8. P. Flores, J. Luo, D. W. Mueller, F. Muecklich, and L. Zea, "Space biofilms - An overview of the morphology of Pseudomonas aeruginosa biofilms grown on silicone and cellulose membranes on board the international space station," (in eng), Biofilm, vol. 7, p. 100182, Jun 2024, doi: 10.1016/j.bioflm.2024.100182.
9. L. Zea et al., "Design of a spaceflight biofilm experiment," Acta Astronaut, vol. 148, pp. 294-300, Jul 2018, doi: 10.1016/j.actaastro.2018.04.039.
10. B. H. Pyle, G.A. McFeters, S.C. Broadaway, C.K. Johnsrud, R.T. Storga, and J. Borkowski, "Bacterial Growth on surfaces and in suspensions. In Biorack on Spacehab," Biological Experiments on Shuttle to Mir Missions 03, 05, and 06. European Space Agency SP-1222, 1999.
11. R. J. McLean, J. M. Cassanto, M. B. Barnes, and J. H. Koo, "Bacterial biofilm formation under microgravity conditions," (in eng), FEMS Microbiol Lett, vol. 195, no. 2, pp. 115-9, Feb 20 2001, doi: 10.1111/j.1574-6968.2001.tb10507.x.
12. W. Kim et al., "Spaceflight promotes biofilm formation by Pseudomonas aeruginosa," (in eng), PLoS One, vol. 8, no. 4, p. e62437, 2013, doi: 10.1371/journal.pone.0062437.
13. P. Flores et al., "Preparation for and performance of a Pseudomonas aeruginosa biofilm experiment on board the International Space Station," Acta Astronautica, vol. 199, pp. 386-400, 2022/10/01/ 2022, doi: https://doi.org/10.1016/j.actaastro.2022.07.015.
14. L. C. Powell et al., "Quantifying the effects of antibiotic treatment on the extracellular polymer network of antimicrobial resistant and sensitive biofilms using multiple particle tracking," npj Biofilms and Microbiomes, vol. 7, no. 1, p. 13, 2021/02/05 2021, doi: 10.1038/s41522-020-00172-6.
15. K. Forier et al., "Transport of nanoparticles in cystic fibrosis sputum and bacterial biofilms by single-particle tracking microscopy," (in eng), Nanomedicine (Lond), vol. 8, no. 6, pp. 935-49, Jun 2013, doi: 10.2217/nnm.12.129.
16. A. Birjiniuk, N. Billings, E. Nance, J. Hanes, K. Ribbeck, and P. S. Doyle, "Single particle tracking reveals spatial and dynamic organization of the E. coli biofilm matrix," (in eng), New J Phys, vol. 16, no. 8, p. 085014, Aug 27 2014, doi: 10.1088/1367-2630/16/8/085014.
17. M. E. Grady, E. Parrish, M. A. Caporizzo, S. C. Seeger, R. J. Composto, and D. M. Eckmann, "Intracellular nanoparticle dynamics affected by cytoskeletal integrity," (in eng), Soft Matter, vol. 13, no. 9, pp. 1873-1880, Mar 1 2017, doi: 10.1039/c6sm02464e.

18. N. Tarantino et al., "TNF and IL-1 exhibit distinct ubiquitin requirements for inducing NEMO–IKK supramolecular structures," Journal of Cell Biology, vol. 204, no. 2, pp. 231-245, 2014, doi: 10.1083/jcb.201307172.
19. J. S. Suk, Q. Xu, N. Kim, J. Hanes, and L. M. Ensign, "PEGylation as a strategy for improving nanoparticle-based drug and gene delivery," (in eng), Adv Drug Deliv Rev, vol. 99, no. Pt A, pp. 28-51, Apr 1 2016, doi: 10.1016/j.addr.2015.09.012.

For Product Safety Concerns and Information please contact our
EU representative GPSR@taylorandfrancis.com Taylor & Francis
Verlag GmbH, Kaufingerstraße 24, 80331 München, Germany